ALBUM OF SCIENCE

*The Biological Sciences
in the
Twentieth Century*

ALBUM

OF

SCIENCE

The Biological Sciences
in the
Twentieth Century

MERRILEY BORELL

I. B. Cohen, General Editor, Albums of Science

CHARLES SCRIBNER'S SONS

New York

0381

1. (*Frontispiece*) Watson, Crick, and DNA. The mid-century revolution in biology provoked by the discovery of the molecular structure of the hereditary substance DNA focused biologists' full attention on biochemical processes in living cells. It marked the rise to prominence of laboratory-based biological research. The American biologist James Watson and the British biophysicist Francis Crick had developed in early 1953 a physical model of DNA that could explain many hereditary and biochemical phenomena in living organisms: they asserted that DNA consisted of two complementary strands twisted into a double helix. The Watson-Crick structure was in accord with all the known chemical facts about DNA, and it suggested how the DNA molecule could replicate itself and transfer genetic information. This picture of Watson (*left*) and Crick was taken in the spring of 1953 in Cambridge University's Cavendish Laboratory. The two are looking at the first full model of DNA.

Copyright © 1989 Macmillan Publishing Company

Library of Congress Cataloging-in-Publication Data
Borell, Merriley.
    The biological sciences in the twentieth century / Merriley
Borell.
        p.   cm. — (Album of science)
    Bibliography: p.
    Includes index.
    ISBN 0-684-16483-3
    1. Biology—History. 2. Biology—Philosophy. 3. Biology—Social
aspects. I. Title. II. Series.
QH305.B67   1988
574'.09'04—dc19   88-14715
                                                    CIP

Previous volumes in the *Album of Science* series:
John E. Murdoch, *Antiquity and the Middle Ages*, ISBN 0-684-15496-X
I. Bernard Cohen, *From Leonardo to Lavoisier, 1450-1800*, ISBN 0-684-15377-7
L. Pearce Williams, *The Nineteenth Century*, ISBN 0-684-15047-6

Published simultaneously in Canada
by Collier Macmillan Canada, Inc.
Copyright under the Berne Convention

1 3 5 7 9 11 13 15 17 19 Q/C 20 18 16 14 12 10 8 6 4 2

Printed in the United States of America

Picture research by John Schultz—PAR/NYC

For
Marion Borell
and
Marv Borell

Contents

Acknowledgments   ix
Foreword, by I. Bernard Cohen   xi
Introduction   xiii

Part I: THE NEW CENTURY
1. *A Vision of Progress*   2
2. *Rise of the Biological Laboratory*   12

Part II: THE EXPERIMENTAL METHOD IN TWENTIETH-CENTURY BIOLOGY
3. *Research Settings*   24
4. *Organisms*   36
5. *Methods of Observation*   46
6. *Modes of Analysis*   56

Part III: TRANSFORMATION OF THE NATURAL HISTORY TRADITION
7. *Paleontology*   68
8. *Embryology*   80
9. *Genetics*   90
10. *Ecology*   102
11. *Evolution*   112

Part IV: INVESTIGATING LIFE PROCESSES
12. *Structure*   124
13. *Function*   136
14. *Integration*   148
15. *Biochemistry*   158
16. *Molecular Biology*   168
17. *Origin and Procreation of Life*   180

Part V: THE HUMAN ANIMAL

18. *Physical Anthropology*   192
19. *Human and Animal Behavior*   202

Part VI: APPLYING BIOLOGY TO SOCIAL NEEDS

20. *Controlling Disease*   214
21. *Preventing Disease*   224
22. *Controlling Population Growth*   232
23. *Improving Agricultural Production*   242
24. *Repairing the Human Body*   252

Part VII: THE NATURE OF THE ORGANISM

25. *The Language of Biology*   264
26. *The New Research Enterprise*   274
27. *Our Place in Nature*   280

Guide to Further Reading   291
Picture Sources and Credits   294
Index   300

Acknowledgments

The growth of the biological sciences in the twentieth century has been multifaceted and complex. The opportunity to present this growth visually in an *Album of Science* has been both exciting and extremely challenging. I am grateful to Bernard Cohen for inviting me to prepare this volume, giving a new direction to the work I began at Harvard University in 1985–1986 through the National Science Foundation's Visiting Professorships for Women Program, for whose research support I am most appreciative.

Many of the ideas developed in this book grew out of that very fruitful year. I owe special thanks to the Department of the History of Science at Harvard and to Everett Mendelsohn, Barbara Rosenkrantz, Hughes Evans, Deborah Coon, Gail Hornstein, Adele Clarke, and Diana Long for encouragement and support in the elaboration and development of these ideas.

The spectacular blossoming of the biological sciences is readily seen in the increasingly specialized work spaces of biologists and in the growing interest of agricultural, medical, and industrial sectors in the fruits of biological research. Pictures convey this fascinating story far more effectively than the words of traditional scholarship.

Selecting the pictures for this volume and exploring their significance has not been a solitary task. Many colleagues provided leads in locating or choosing suitable illustrations. Toby Appel directed me to numerous fine photographs in the archives of the American Physiological Society. Peggy Kidwell introduced me to the collections of the Smithsonian Institution. William Schupbach and Renate Burgess helped locate materials at the Wellcome Institute for the History of Medicine in London. Maida Goodwin sent valuable pictures from the Smith College Archives. Garland Allen supplied much illustrative material; C. Stewart Gillmor was also a useful source. John Schultz and my editor, Richard Hantula, secured illustrations for me from numerous collections throughout the world. I owe a particular debt of gratitude to Robert Pool and Gene Cittadino for their advice and enthusiasm in guiding me to valuable scientific and historical sources.

Other colleagues read and commented on various drafts of the manuscript. I especially wish to thank Garland Allen, Jane Maienschein, Pnina Abir-Am, Diana Long, and Alan Goodman, who each provided very helpful and timely comments on significant portions of the text. Lily Kay and Charles Kochakian kindly shared their unpublished manuscripts on instrumentation and twentieth-century research techniques with me.

Zelda Haber, Joan Gampert, Marvin Friedman, Emil Chendea, Mynette Green, and staff at Macmillan were responsible for design, layout, and key production work; my thanks to all.

Throughout this lengthy project Richard Hantula never failed to offer his excellent editorial assistance, good humor, and steady encouragement. I am grateful both to him and to my companion and colleague Robert Pool for continual support in the conception and realization of this book.

*F*oreword

This volume is the fourth to appear in the *Album of Science* series. The goal of this series is to provide a pictorial record of the growth of the scientific enterprise, displaying in images what science has been like and conveying a sense of the perception of science by men and women, both scientists and nonscientists, living in different times. Accordingly, the present volume is not a mere record of great discoveries, although of course the most important advances in the life sciences are represented here. In fact, it would not be possible to produce a complete pictorial record of the "great discoveries" in the framework of the *Album of Science,* since many important discoveries are of an abstract nature and do not readily lend themselves to graphic images.

The present volume, like its companion on the physical sciences, differs from the three earlier volumes—*Antiquity and the Middle Ages* (by John E. Murdoch), *From Leonardo to Lavoisier* (by myself), and *The Nineteenth Century* (by L. Pearce Williams)—in that the whole time span has been recorded for us by photography. Accordingly, most of the illustrations are photographs, although there are some reproductions of manuscript pages and of famous publications, such as James Watson and Francis Crick's announcement in *Nature* of the double helix structure of DNA. The photographs include a number made with the optical or electron microscope that are of notable beauty. There are also many drawings and graphic representations of cells, tissues, and other biological entities and phenomena.

The book has few portraits in the formal sense, since the aim of the series has been to show what science was like, rather than to display the visages of famous men or women. But we do see scientists working in their laboratories or in groups. Indeed, the volume shows the whole environment of the biological sciences in our times. Thus, considerable attention is devoted to layouts and floor plans of laboratories in which important science was done, pictures of key institutions (such as the Marine Biological Laboratory at Woods Hole, Massachusetts), as well as scenes from individual laboratories showing the way in which scientific results were being produced.

One of the most impressive aspects of this collection is that the photographs have been drawn from a wide variety of archival sources; many images have never before been reproduced in book form. Collectively, they give the viewer the experience of actually "having been there" when great discoveries were in the process of being made.

Of course, even the most successful photographs of scientific institutions, laboratories, experiments, or other aspects of research do not speak for themselves. Pictures recording the development of science require interpretation. Merriley Borell provides a text that takes us on a visual tour of the panoramic development of the life sciences, explaining important concepts, events, and developments and indicating their scientific and social significance.

An especially attractive feature of Merriley Borell's volume is her constant attention to the social matrix and social implications of scientific research and discovery. She depicts the international character of the scientific enterprise, the growth of "big science" and its consequences, and the relations between so-called pure or disinterested scientific research—the search for truth—and the needs of society. She stresses the important and often unanticipated consequences that arise with the application of advances in fundamental knowledge.

The history of the biological sciences in the twentieth century reveals a complex network of causal interactions between scientific research and its ultimate effects. Consider scientists' progress in understanding the nature of growth processes in plants. A key development was the discovery of auxins, growth substances produced in the tips of plants. Work on this topic goes back at least to studies by Charles Darwin and his son Francis on the apparent tendency of plants to bend toward a source of light. These late-nineteenth-century experiments were the beginning of a considerable body of research in which, culminating in the work of Frits Went and Kenneth V. Thimann, the first growth substance was identified, isolated, and chemically analyzed. This was made possible, in part, by the invention in the 1930s of the microbalance and of techniques for dealing with very tiny quantities of material. The identification of auxin provided a chemical answer to Darwin's research problem: if an oat shoot is exposed to a source of light, auxin will move away from the side exposed to light; the result will be a far greater rate of growth on the shady side, pushing the shoot toward the light.

Before long, it was found that such chemical substances controlling growth in plants could be put to use in ways that have deep and far-ranging economic and social consequences. Spraying crops with auxins made it possible to improve harvests (for instance, by keeping apples on the tree a little longer to get fuller growth), to preserve food longer under storage conditions, and eventually—in perhaps the most popular and widespread application in use in the decades following World War II—to control the growth of weeds. This last application held out great promise both for home gardeners wishing to free their lawns of certain major types of weeds (primarily broadleaf annuals) and for commercial farmers wishing to increase productivity by chemical weeding in their fields. During the Vietnam war synthetically produced auxins (primarily 2,4-D and 2,4,5-T) were used for crop destruction and tree defoliation, constituting a major component of "Agent Orange," in which contaminants were later found to have very harmful effects on both soldiers and native civilians. This destruction of fertile lands was a social consequence totally unforeseen in the first decades of peaceful applications of auxins (chiefly 2,4-D), although it rather soon became evident that accumulation of auxins had the potential of being harmful to animal life as well as to human beings. This complex series of events demonstrates dramatically the ways in which, often quite unexpectedly, the fruits of scientific discovery may be used for what seem to be beneficial or destructive purposes.

The story of auxins illustrates also the dangers of large-scale injection of chemical substances into the environment. On the one hand, new techniques—from the hybridization experiments of the early part of the century to recent innovations based on genetic engineering—have the potential to increase the food supply and to help us prevent and conquer important diseases. On the other hand, the indiscriminate use of chemicals—herbicides, insecticides, fertilizers, and air sprays—poses dangers to the environment that will affect not only our own lives but those of future generations.

A particular virtue of Merriley Borell's volume is that it makes us conscious of this dual character of the biological sciences in the twentieth century. The century has brought advances in knowledge of life processes but has also brought important applications that have the power both to improve the quality of life and ultimately to impoverish our environmental habitation. Making decisions about the potentialities of science requires education. Merriley Borell's volume will not solve these problems for us, but it will enrich our understanding of the progress of science and will thereby increase our understanding of the nature of the scientific process. No reader can come away from this volume without a deeper appreciation of the development of the life sciences as a heroic intellectual enterprise within our society, an enterprise that ultimately—through its applications—affects the lives of every inhabitant of our planet.

I. Bernard Cohen

Introduction

The twentieth century has seen biology emerge as an important intellectual and social force. The elaboration of the theory of evolution, the development of effective drugs and powerful pesticides, the analysis of the biological basis of human behavior, have challenged traditional explanations of our lives and our place in nature. Advances in the life sciences have revolutionized medicine and agriculture and given rise to new industries based on biological products and biochemical processes. As a result, biologists have claimed new power to mold and shape the world of the future. Yet the diverse ways in which biological theories have been employed in the past to alter both human life and nature suggest that there are many alternative options for the future and that we ought to reflect with some care on these choices.

At the center of the story of the life sciences' rise to prominence in the twentieth century is the biological laboratory. Early in the century the laboratory replaced the field as the center of research activity. It became the premier site for the development of new and specifically quantitative techniques for analyzing life processes. The increasingly important role of the laboratory in research and the resulting structural reorganization of biology were elements of the general transformation from "little science" to "big science" that took place in the sciences generally—a transformation that accelerated after World War II with the infusion of vast sums of money into research. In biology, fundamental changes were evident already at the beginning of the century, both in approach (the new focus on the experimental) and in organizational structure (the industrializing processes involved in the shift from little science to big science). Such changes affected the selection and development of research problems in disciplines as diverse as physiology, genetics, evolutionary theory, psychology, and, ultimately, molecular biology.

Despite the separate historical roots and traditions of botany and zoology on the one hand and medically oriented functional studies on the other, biology in the twentieth century came to be unified both by method and concept: the experimental method of analysis gained ascendancy, and the theory of evolution was applied to all organisms. Plants and animals (and minerals) had long been studied in a descriptive way as a part of natural history, but around the turn of the century the historically separate botany-zoology and anatomy-physiology traditions merged into a coherent experimentally oriented biology that explained form, function, and change over time. The "life sciences" or "biological sciences" of the title of this volume represent diversification of this newly developed, self-consciously experimental biology as it expanded in the twenties, thirties, and forties, encompassing accessible problems in embryology, genetics, bacteriology, biochemistry, and cell physiology.

The discovery around mid-century that DNA—deoxyribonucleic acid—is the genetic material consolidated the experimental emphasis and gave additional impetus to the institutional changes. Scientists quickly reformulated many long-standing research problems as they sought to reduce complex biological phenomena to an orderly sequence of molecular events. The holistic emphases of the first several decades receded as reductionist experimental techniques drew attention to individual molecules within the cell. With the emergence of molecular biology in the 1950s, biochemistry and molecular biology, cell and developmental biology, and organismic and evolutionary biology separated into distinct domains. The neurosciences may well produce another such conceptual and institutional reorganization of the life sciences in the near future.

Biology has made its mark in many areas of human endeavor in the twentieth century—in medicine, agriculture, industry, and philosophy. Alongside science's increasing ability to control disease, population growth, and human development, however, continual human intervention in natural processes has raised troubling questions. This intervention has produced unpredictable and serious side effects, especially in the environment. Thus, the final social impact of the century's biological discoveries has yet to be assessed.

This volume of the *Album of Science* surveys the rise and transformation of the biological sciences in the twentieth century—from the organism-centered field observations of the naturalist to the molecule-oriented analyses of the laboratory investigator. It is both a visual portrayal of science as a human activity and a historical interpretation of biology's changing structure and content. The volume is divided into seven parts that follow the approximate chronological development of each of the major biological disciplines while linking together conceptually related subject matter. Thus, evolutionary issues reappear in several chapters as the natural history tradition of the early twentieth century (Paleontology, Ecology, Evolution) becomes tied to functional questions in mid-century studies (Molecular Biology, Origin and Procreation of Life). The book unfolds in three separate layers: the social and organizational structure surrounding the experimental approach (Parts I-II), the evolving content of the life sciences themselves (Parts III-V), and finally the social impact of the new insights into life processes (Parts VI-VII).

Part I begins by exploring the vision of progress that heralded the twentieth century. The biological laboratory is then introduced as it grew early in the century from a mere set of tables with a few microscopes to a complex, mechanized work area. Part II reviews the major methodological themes of twentieth-century biology—research environments, organisms, and techniques of observation and analysis. Parts III and IV follow the research themes of the major subdivisions of biology, the natural history tradition and functional studies, respectively. Part V reviews the impact of these related investigations on understanding of the human species as expressed in anthropological and behavioral studies. Part VI examines the social impact of recent biological research in medicine, public health, family planning, agriculture, biotechnology, and biomedical engineering. Part VII probes the significance of twentieth-century developments for our understanding of the natural world and the place of *Homo sapiens* within it.

The pictures selected for this volume suggest critical episodes in the growth of the now vast enterprise of the biological sciences, as well as foci for further study. The book's images show the immense range of illustrations and graphic material used by scientists and the popular press in the twentieth century. Ordinary photographs, optical and electron micrographs, drawings from scientific works, paintings, sketches, and caricatures portray the rich pageant of biological discovery. In addition to documenting the spectacular rise of the biological sciences, these images reveal the human effects of modern science: from early public perceptions of X rays to imagined automatons, to environments altered by new breeds of plants or chemical effluents. These pictures are but a sampling of the wealth of visual evidence yet to be explored by historians and sociologists of science probing the complex relationships between science and society; indeed, critical historical analysis of most of the events portrayed in these pages has only just begun.

Readers who may wish to explore further the issues and events depicted here should turn to the Guide to Further Reading at the end of the book.

Part One

THE
NEW
CENTURY

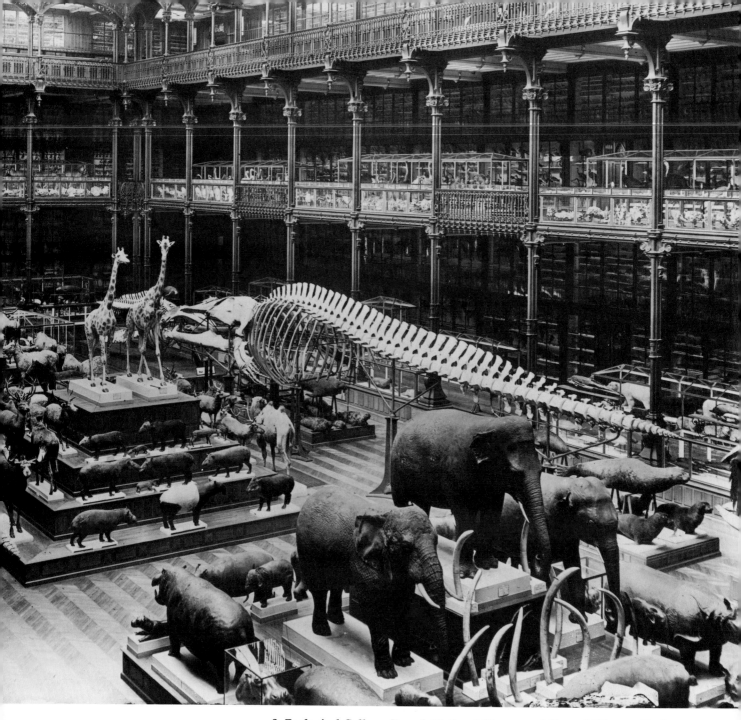

2. Zoological Gallery, French National Museum of Natural History, Paris.
Skeletons and stuffed specimens arrayed across the main floor of this spacious hall and on additional levels at the sides display the diversity of the animal kingdom. The mammals shown here, collected in part from expeditions outside Europe, illustrate the grandeur and variety of animal life.
The photograph was taken about 1890, not long after the exhibit area was inaugurated. The gallery remained open to the public until 1965.

1

A Vision of Progress

In the nineteenth century the science of biology entered the public eye, challenging traditional philosophy and religion and provoking endless debate over the place of human beings in nature. By 1900, scientists' voyages of exploration, new discoveries of fossils, and Charles Darwin's evolutionary synthesis had captured the popular imagination. Halls of diverse creatures formed the basis of natural history collections in major museums across the world—many established in the preceding decades. While vigorous debate raged over the ultimate meaning of this diversity, the sheer size and magnificence of a brontosaurus or a mammoth were a cause for public wonder. Housed in cathedral-like settings, the endless arrays of specimens urged reflection on the history of the natural world and reassessment of the philosophical foundations of Western thought. Science was to take a pivotal role in the new world of the twentieth century. In biology, evolution would be the focal philosophical issue.

Darwin's evolutionary hypothesis was one of the crowning achievements of nineteenth-century biology. Darwin argued that organic nature changed gradually over time, and he suggested a mechanism for how this change might occur. His speculations and predictions underlay subsequent work in the natural history of plants and animals, the field of research that was to blossom into the life sciences in the twentieth century. His theory also provided the interpretive context for much twentieth-century research on biological function,

research that had derived its initial impetus from the application of new precision instruments in the nineteenth century. In short, evolutionary theory—along with newly fruitful chemical and microscopic research on the activities of living things, as well as wide adoption of the methods of the physical sciences—shaped and structured further experimental analysis of life processes.

A second major achievement of nineteenth-century biology was cell theory—the recognition that organisms consist of basic units called cells, which propagate by dividing. Based on the synthetic work of the botanist Mathias Schleiden and the zoologist Theodor Schwann as extended by the pathologist Rudolf Virchow, the concept of the cell as the basic unit of life had gained wide currency by 1900. Meanwhile, the rising science of organic chemistry furnished chemical analyses of fundamental life processes like respiration and fermentation. These parallel developments encouraged further study of the metabolic activities of tissues and cells.

A seminal figure in nineteenth-century biochemical research was the great French chemist Louis Pasteur. Pasteur focused the attention of biological scientists on the chemical activities of a multitude of microscopic creatures responsible for fermentation and disease. These "germs" became the subject matter of the prolific new fields of bacteriology and microbiology. Already in the nineteenth century bacteriological research profoundly affected such diverse areas of activity as medicine and surgery, agriculture, the beer and wine industries, and the manufacture of pharmaceuticals. Microorganism research signaled vast possibilities for the prevention and treatment of disease, the improvement of agricultural production, and the development of chemically based industry. These opportunities would accelerate in the twentieth century, as biological insights permeated industry and commerce.

Already in the nineteenth century scientific research, especially that of practical import, had become a cause for intense national pride. Prominent scientists were publicly feted. Examples late in the century were Pasteur and the British surgeon Joseph Lister, developer of the method of antiseptic surgery. Medicine and surgery were fundamentally altered by these two. Their successors, among them the German medical scientists Robert Koch and Paul Ehrlich, and the Russian émigré Élie Metchnikoff, laid the foundations for several new medical specialties that would give rise to a scientific therapeutics in the twentieth century.

Research laboratories, the sites of the new discoveries, multiplied. Research institutes composed of many laboratories were often publicly endowed; in the late nineteenth century several were founded as national centers for the study of the various biological processes observed in disease. Institutions like the Lister Institute in London and the numerous Pasteur Institutes that emerged throughout the world combined the interests of basic biological and medical researchers with the goals of a growing pharmaceutical industry that sought to provide new and powerful drugs for medicine.

The pharmaceutical industry flourished in the late nineteenth century, reflecting technical advances in chemistry. It produced new biological products like antitoxic sera, vaccines, and extracts harvested from animal tissues and cultures of microorganisms, as well as traditional purified plant drugs. The new century quickly brought new techniques for obtaining these products (which acquired the name "biologicals"), for ascertaining their composition, and for isolating and, ultimately, synthesizing other, more powerful, specifically antibiotic substances. Research laboratories featured prominently in these discoveries and in this growth, opening up opportunities for the analysis, alteration, and even control of life processes, especially by chemical manipulation.

Faith in the power of science to provide technical solutions to human problems emerged as a dominant creed early in the new century. This faith found expression in virtually every area of biological investigation, with many life science researchers committed to the task of solving human problems. At this time, belief in the use of eugenic principles to guide human breeding gained a wide following; later in the century, it became possible to replace malfunctioning human organs with mechanical organs or transplanted natural organs. Physicians expected to be able to cure and even eliminate disease. As biological researchers, aspiring to meet these goals, embraced the methods of chemistry and physics (while not abandoning traditional observational techniques), they gained social power and prestige as well as new authority and responsibility in shaping human welfare. The products of their labors—the techniques they developed, the discoveries they announced, even the assumptions of their scientific method—profoundly affected social and economic development, yielding revolutionary changes in biology, medicine, agriculture, and industry.

3. Central Hall, British Museum (Natural History), London. Visitors entering the museum around the turn of the century saw this vista. Prominent in the Central Hall were statues and busts of museum officials and contributors to the museum's collections, including Sir Richard Owen, Professor Thomas Henry Huxley, and Sir W. H. Flower. The display cases illustrated major points of interest in natural history and zoology, subjects then in the process of being transformed by Darwin's theory of evolution. The building was erected between 1873 and 1880.

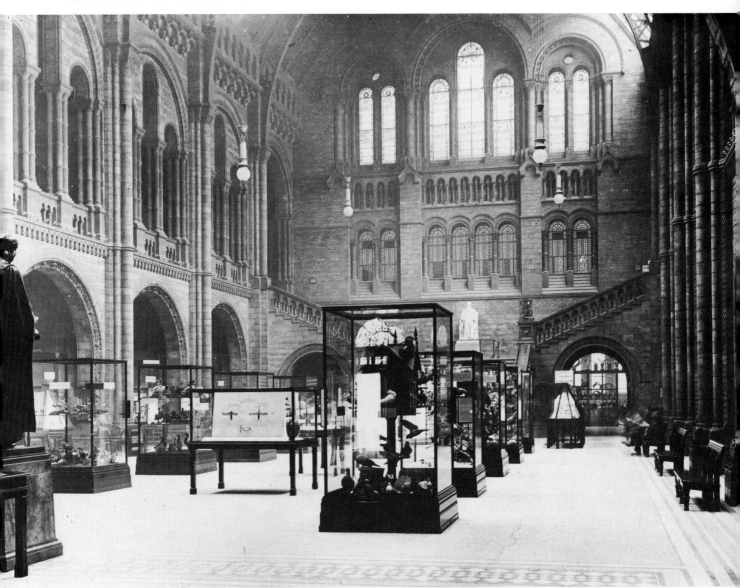

4. Bird Hall, British Museum (Natural History). Here, individual birds are exhibited in their natural surroundings, often with nests, eggs, or young. The cases of stuffed specimens illustrate not only the variety of form and plumage found among birds in nature but also the principles of evolution. Important evidence for the theory of evolution derived from Charles Darwin's observation of finches in the Galápagos Islands in the 1840s.

5. Natural History Museum, Berlin. Whale skeletons dominate this view of mammalian specimens in the museum's Central Hall. Note the skylight illuminating the space, a common feature of museum design in the late nineteenth and early twentieth centuries.

6. Darwin. This unusual, contemplative portrait juxtaposes Darwin with the microscope. While the microscope was very important to the development of nineteenth-century biology generally, the particular contributions of the naturalist Darwin derived principally from his field observations and from his synthesis of data from subjects as diverse as geology, zoology, botany, and embryology. The twentieth century would see the gradual unification of natural history and laboratory research, partly as a result of new research emphases prompted by Darwin's speculations.

7. Drawing from Pasteur's research on fermentation. In the mid-nineteenth century it was believed that yeast was just a chemical catalyst in alcoholic fermentation, but Louis Pasteur argued that yeast was a living organism, whose biological activity turned sugar into alcohol. Different organisms ("germs") produced different substances by fermentation. Pasteur's insight led to recognition of the microscopic agents of disease.

8. Page from Pasteur's manuscript. By the end of the nineteenth century Pasteur's work had stimulated research in bacteriology, biochemistry, and immunology, resulting in the isolation and identification of the causative agents of major infectious diseases. This led to strong expectations that science would soon be able to control all disease.

9. Pasteur's seventieth birthday, 1892. At a jubilee celebration at the Sorbonne in Paris, the frail Pasteur (*center*) was assisted by Sadi Carnot, president of the Third French Republic.

10. Ehrlich in his office. Scientific research proceeds amid stacks of papers and journals as well as in the laboratory. The German medical scientist Paul Ehrlich shared the Nobel Prize in physiology or medicine for 1908 with the Russian-born Élie Metchnikoff. Building on the work of Pasteur, Ehrlich and Metchnikoff laid the foundations for twentieth-century studies of immunity.

11. Koch in the laboratory. The German bacteriologist Robert Koch received the Nobel Prize in physiology or medicine in 1905 for the discovery and isolation of the tubercle bacillus. This wood engraving shows Koch in 1890, the year he announced the discovery of tuberculin, a substance produced by the bacillus, as a cure for tuberculosis. Although not actually a cure, tuberculin later proved to be of value in diagnosis. Note the tools of research: the microscope, the chemical reagents, and the animals (in cages).

12. World's Columbian Exposition: Electricity Building. The Columbian Exposition, held in Chicago in 1893, showcased the nineteenth century's achievements in technology and science. One of the greatest technological wonders of the age was electric power. A new alternating-current electrical system, developed by the Croatian-born inventor Nikola Tesla, was used to light the exposition. The fair's educational exhibits displayed scientific instruments. Electrification and newly developed precision apparatus were coming to both research and teaching laboratories. In the new century mechanical equipment would be replaced by electrical and electronic devices.

13. Chemical technology at the turn of the century.
An 1895 exhibition at the Royal College of Surgeons
in London was the site of this exhibit by the British
chemical and pharmaceutical manufacturer Burroughs
Wellcome & Co. Note the variety of products on display. Expectations of rapid progress in science were
widespread at the time. In the twentieth century the
development of potent new drugs would dramatically
alter physicians' abilities both to control and to cure
disease.

14. Turin exposition. This part of the Burroughs
Wellcome exhibit at the 1911 International Exposition
of Industry and Labor in Turin, Italy, focused on the
preparation of the alkaloid ergotoxine, whose medical
uses included the stimulation of labor (during child-
birth) and milk ejection. The display shows some of
the extraction procedures underlying drug manufacture that were to be so successfully exploited in the
new century.

15. Roscoff Biological Station, France. Henri de Lacaze-Duthiers, a professor at the Sorbonne, established a "Laboratory of Experimental Zoology" at Roscoff, Brittany, in the 1870s. This view of the research aquaria room dates from about 1890. In the front, at left, is Yves Delage, who succeeded Lacaze-Duthiers as director in 1900. Delage's research focused on problems in marine zoology, as well as general biology.

2

*R*ise of the *B*iological *L*aboratory

As the twentieth century opened, the life sciences were full of promise. The observational data of earlier years, the explorers' finds, the theorists' speculations, had established an intriguing set of questions for further research. The rise of new industries based on biological products fed public enthusiasm for this endeavor—enthusiasm reflected in support for museums. Much remained to be learned, and biologists increasingly followed the lead of the physicists and chemists by turning to measurement and experiment. The new century consequently saw a widespread expansion in the everyday work space of biological scientists.

Early in the century spacious, carefully planned, fully equipped laboratories rapidly replaced small, badly lit workrooms consisting of only a few tables and a microscope. Educational institutions and indeed nations competed with one another in establishing new research facilities. These amenities had begun to appear with increasing frequency in the life sciences in the latter half of the nineteenth century. Such laboratories were fully developed distinctive social institutions in both Europe and North America by the 1920s.

As simple work spaces gave way to carefully planned areas housing elaborate mechanical and electrical equipment, research institutes grew to comprise clusters of many related laboratories, where the structures and life processes of organisms could be thoroughly studied. These facilities developed very rapidly in physiology, where impressive recording instruments based on the

revolving kymograph drum recorder made extensive quantitative study possible. Bacteriological laboratories also blossomed. Spreading from Germany and France to elsewhere in Europe and to Asia and North America, research laboratories became a fundamental feature not only of physiology and bacteriology but of all the emerging life science disciplines by the turn of the twentieth century. In particular, the medicine-related sciences, especially microbiology and biological chemistry, showed this rapid institutional expansion.

In zoology and botany (the traditional fields of natural history from which grew the multifaceted biological sciences) the changing emphasis could be seen in the development of marine laboratories, like the Roscoff Biological Station in France or the Naples Zoological Station in Italy. Here embryologists took advantage of the experimental methods first advocated by physiologists to study the simpler forms of life—animals such as sea urchins, marine worms, and starfish. At the marine laboratories, organisms caught and collected fresh from the seashore were kept in aquaria and observed or used for experiments. Initially occupied with microscopical studies, marine laboratories gradually acquired the specialized recording instruments and chemical apparatus that had come to characterize experimental research. As time passed, the growing focus on the laboratory redefined and enhanced natural history, in the sense of the systematic study of the life histories and habits of animals and plants. There emerged a new research emphasis called "experimental biology"—a catchword in the first third of the twentieth century.

As experimental manipulation came to supplement, even at times supplant, traditional observational techniques, special research spaces or work stations were created within individual laboratories to make use of the combined techniques of microscopy, chemistry, and physiological recording. The general electrification of laboratories and physiological instrumentation early in the century extended biologists' working hours and made it possible to introduce new equipment for studying physiological processes in living things. Researchers needed special rooms for their equipment, and they often hired skilled mechanics to keep the apparatus in good condition. In addition, experimenters developed special quarters for research animals—such animal "colonies" ensured a supply of healthy, genetically similar animals whose physiological condition could be controlled by the investigator.

As the work space, status, and role of the experimenter grew, so did the use of laboratory work in teaching science. Small workrooms used by only the professor and a few advanced students gave way to large work spaces where multitudes of students could receive hands-on training. In the United States, by the early twentieth century even smaller colleges had begun to provide training by the "laboratory method." As teaching techniques were routinized, and as apparatus became standardized (and produced more cheaply by a growing industry of laboratory instrument manufacturers), even relatively small institutions could afford to teach biological science to large numbers of students. This institutional transformation, which had first taken hold in chemistry and physics, spread rapidly in the life sciences early in the century, just as newly applied analytical techniques from chemistry and physics were yielding exciting and important results in biology.

By the 1920s the main features of this institutional transformation were in place: large numbers of students were being introduced to experimental research techniques not only by lecture but also by regular practical experience in specialized teaching laboratories. These large student work spaces looked very much like factories arranged for the production of scientific knowledge and for the production of skilled scientific workers.

Indeed, industrializing processes had begun to penetrate biological research. Here were the origins of the transition from "little science" (exemplified by the solitary researcher) to "big science" (teams of researchers working with expensive hardware in specially designed laboratories)—a process that would accelerate spectacularly in all the natural sciences after World War II. The rising pharmaceutical industry needed large numbers of laboratory workers. So did agriculture, where scientific study of nutrition, genetics, and reproduction in the first half of the century was to have a marked effect on cash crops like corn and wheat and on the marketability of stock animals. Medicine, too, was revolutionized, incorporating new technologies and experiencing a very rapid transformation into a high-status profession. Physicians and surgeons were able to save lives with new wonder drugs and new therapeutic techniques previously unthinkable.

Thus, the twentieth-century life sciences embraced the experimental laboratory. As biological science became a socially important activity, the observations of naturalists gradually ceded their predominance to the probing experimental manipulations of laboratory researchers. Biologists achieved new understanding and control over nature by joining physicists and chemists at the laboratory bench. The rise of the biological laboratory early in the century—and its transformation and expansion in the middle decades—reflects both the new interests of biologists and the shaping of the new interests to meet perceived social needs. As these early aspirations matured, they bore fruit in the creation of potent new drugs, the breeding of high-yield hybrids, and the development of a technologically oriented health care industry.

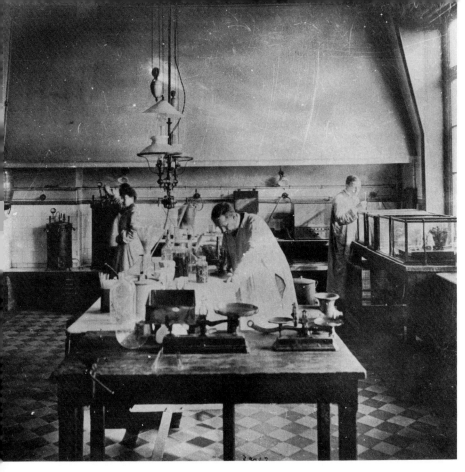

16. Laboratory interior, Pasteur Institute, Paris. This laboratory was used by Élie Metchnikoff, who discovered phagocytosis, a basic form of defense of the body's immune system against foreign organisms. Metchnikoff joined the institute in 1888. The available equipment included a sterilizer (*rear left*), a microscope (*center*), and balances (*front*).

17. String galvanometer room, Harvard Medical School. Galvanometers in this research room were used to detect minute electric currents in the body. The room, photographed early in the twentieth century, reflects the growing use of electrical apparatus (note the numerous wires across the ceiling) and the increasing specialization of equipment and research space.

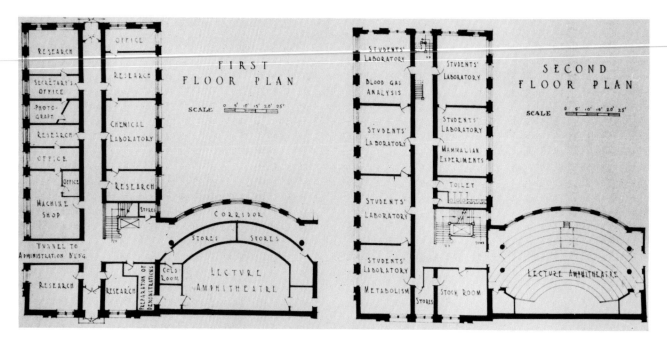

18. Physiological laboratories, Harvard Medical School. Here is more evidence of the increasing differentiation of laboratory space in the early part of the century. These floor plans reflect an increasing compartmentalization of research and teaching activities, along with the organization of specialized support functions—note the machine shop, stockrooms, and animal rooms. Laboratories for chemical analysis, instrument rooms for electrical studies, and animal operating rooms may be seen alongside rooms specially fitted for blood gas, metabolic, and mammalian research.

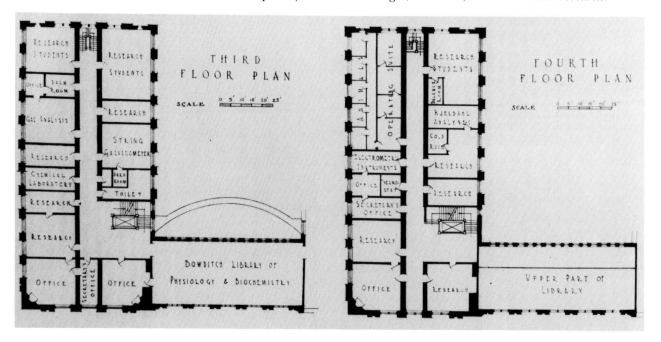

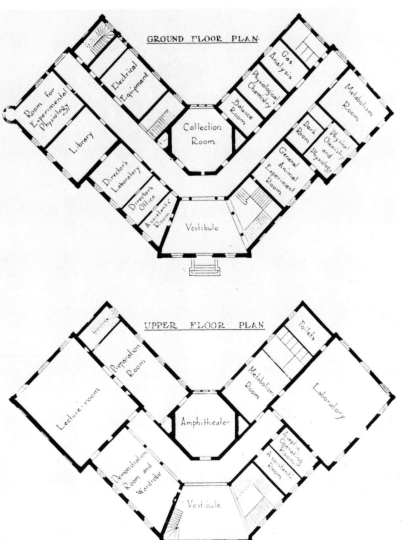

GROUND FLOOR PLAN

Room for Experimental Physiology

Library

Director's Laboratory

Director's Office

Assistant's Room

Electrical Equipment

Collection Room

Vestibule

Gas Analysis

Physiological Chemistry

Balance Room

Dark Room

Physical Chemistry and Physiology

General Animal Experiment Room

Metabolism Room

UPPER FLOOR PLAN

Lecture-room

Preparation Room

Demonstration Room and Wardrobe

Amphitheater

Vestibule

Metabolism Room

Aseptic Operating Room

Assistant's Room

Toilets

Laboratory

19. Institute of Physiology, University of Bern, Switzerland. Called the Hallerianum after the eighteenth-century Swiss physiologist Albrecht von Haller, this building was constructed between 1892 and 1894. The ground floor was devoted to research, while the upper floor was designed for instruction. The institute also included a basement that in the 1920s contained a workshop, a machine room with electrical equipment, a frog cellar, and special rooms for large equipment like the string galvanometer and calorimeter.

20. Room at the Bern institute. This individual research room, photographed in the 1920s, features specialized apparatus for micro-gas analysis and tonometric determinations. It is also outfitted for the examination of animal organs and the chemical work typical of physiological research of the period. Commercially made apparatus was gradually replacing individually crafted units, as scientific supply companies became an important support industry for research.

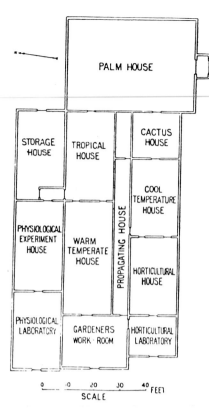

21. Lyman Plant House, 1902. Separating greenhouse and laboratory functions in space designed for student instruction, this Smith College structure reflected the increasing importance paid to plant physiology as experimental methods gained prominence.

22. Botany class. A 1904 photograph of students at work in the Lyman Plant House.

23. Greenhouse experiment room. Early in the century plant physiology studies used worktables and measurement apparatus such as these at Smith College. The tables were designed for the "convenience of working [while] standing." Hands-on experimental work became an important component of educational reform in the United States in that period.

24. Laboratory, the Sorbonne. This view of part of the Sorbonne's physiology facilities dates from about the end of the nineteenth century. Both women and men here explored the newly developing techniques of biological chemistry, a subject that was to burgeon in later decades.

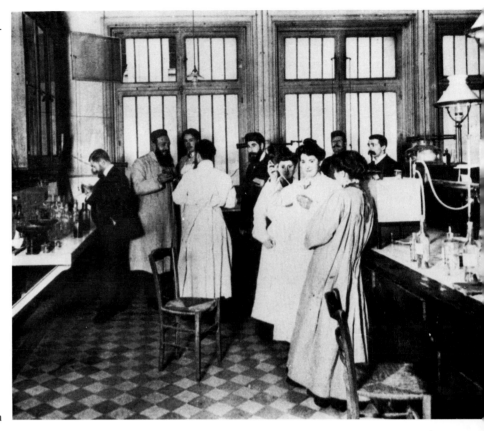

25. Laboratory, Pasteur Institute. Another French teaching laboratory, this one was photographed in the first decade of the new century. The students have microscopes at their disposal. Overhead, "fever charts" show the characteristic patterns of temperature variation associated with recurring fever. An enlarged image of the appearance of the blood in recurrent fever is mounted amid the graphs.

26. Elementary histology room, Cambridge University. Factory-like in appearance, the large hall and its galleries could accommodate 150 students, who worked simultaneously, studying microscope slides. Light was provided by the windows and skylights as well as by electric lamps used in "dull weather."

27. Elementary experimental laboratory, Cambridge. Here is another early-in-the-century view of the facilities of the Cambridge Physiology Department. The rows of kymographs, or "wave writers," recorded physiological events. The revolving drums were driven by a shaft connected to an electric motor. Sixty-two pairs of students could work in the room. Production of kymographs and other apparatus for student use was becoming an important support industry by the end of the nineteenth century.

28. Lecture Hall, Faculty of Pharmacy, Paris. The twentieth century saw increasingly large numbers of students accommodated in specially designed lecture theaters equipped with demonstration tables for presenting experiments. Gas, water, and electrical fittings became an essential feature of teaching-lab design.

29. Practical lab work at the Faculty of Pharmacy. Emphasis on hands-on experimental work for students had begun in chemistry, spread to physics, and then, during the late nineteenth century, entered biology. In the twentieth century all branches of the life sciences adopted the experimental method.

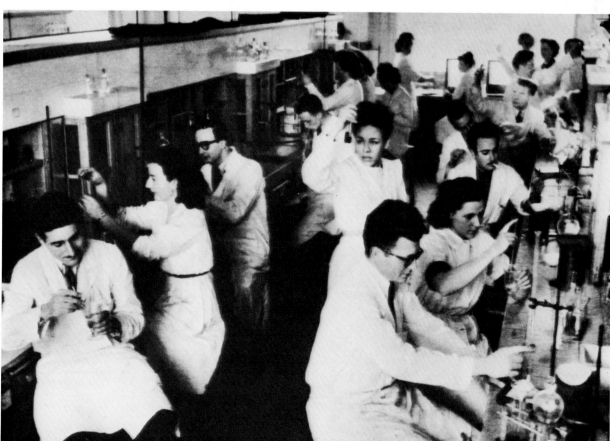

Part Two

THE EXPERIMENTAL METHOD IN TWENTIETH-CENTURY BIOLOGY

30. Collecting specimens. Here, an 1895 botany class from the Marine Biological Laboratory at Woods Hole, Massachusetts, visits Tarpaulin Cove at low tide. From the late nineteenth century, seaside laboratories provided abundant opportunities for collecting and studying plants and animals. Botanists and zoologists at that time were especially interested in studying the embryological development of organisms and in classifying the diverse forms of marine life. Specimens were carried back to the laboratory for observation and dissection, as well as for study under the optical microscope. Analysis of the life histories, development, and distribution of marine organisms constituted the major focus for field courses taught at marine biological stations of the period.

3

Research Settings

Collecting nets, shovels, boxes, pails, and jars at the ready, biologists gather specimens wherever organisms are found. Of course, not all subjects of study can be taken home for examination. The twentieth century afforded research sites ranging from tropical rain forests to deserts, from beneath the sea into space. While evolutionary theory and the study of diversity in nature remained the unifying theme, experiment emerged as the dominant mode of analysis. Exploration and collection came to be supplemented by routine measurement of environmental and physiological variables.

Seeking to study plants and animals in natural settings, biologists increasingly moved their equipment and laboratories into new environments. As they brought instruments and analytical apparatus out into the field, they put together temporary laboratories. Where organisms were plentiful, researchers built field stations and, eventually, elaborate laboratory facilities. An early example was the Naples Zoological Station in Italy; it was founded in the 1870s. By 1900, marine biological laboratories, many modeled on the Naples station, had been established in such countries as France, Great Britain, Monaco, the Netherlands, Norway, Australia, Japan, and the United States (notably the Marine Biological Laboratory at Woods Hole, Massachusetts). Later, researchers built biological field stations at inland sites as well. In the South American jungles, for example, tropical organisms and diseases could be studied *in situ*. Some of these stations became permanent institutions. Others

were merely makeshift temporary structures like tents and shacks, where the researcher might have a microscope, a centrifuge, and perhaps a sterilizer to help in obtaining and analyzing fresh specimens.

By the early years of the century most European nations had set up major research institutes for the study of the medical sciences. It was noted in Chapter 1 that Pasteur Institutes, modeled on the one established in Paris in 1888, sprang up throughout the world. The extensive facilities of the Kaiser Wilhelm Institute for Medical Research at Heidelberg, completed in 1930 and one of several related scientific research institutes in Germany, contained both residential quarters and animal pens. By mid-century, institutions such as these had grown into huge research centers employing hundreds of scientists engaged in a wide variety of research projects. Large sums of money supported these endeavors. The money most often came from foundations or governments, rather than, as in earlier eras, individual patrons of science.

Biologists, like other natural scientists, grew to depend on foundation and government grants for their livelihood. Grants were used not only to support researchers but also to purchase chemicals and instruments or to maintain large animal colonies or experimental farms for research. In the United States and Great Britain routine government support for basic biological research in academia did not become common until after

World War II, although both agricultural and medical research had long been supported by the government or private foundations. (The American network of agricultural field stations for experimental research was established in the late nineteenth century.) German research laboratories were state supported already in the late nineteenth century.

Most of these marine, agricultural, medical, and other biological research institutions produced distinguished work focused on a specific area or areas—for example, bacteriology, embryology, or genetics. Scientists wishing to learn the latest analytical techniques or to study certain organisms traveled to well-known stations or institutes. One such, the Roscoff Biological Station, which had been founded in 1872, grew by 1987 to include fifty-five research labs in biochemistry, electrophysiology, molecular biology, and oceanography, three research vessels, libraries, conference rooms, classrooms, a computer center, residences, and a public aquarium and museum.

In the latter decades of the century researchers placed elaborately equipped facilities at great depths in the sea and in space. In these artificial environments biological scientists studied human responses to changing pressures and/or weightlessness and monitored their own physiological activities through electronic recording devices. They conducted experiments, evaluating how plant growth might be affected by conditions in orbit, say, or how small animals might react to space flight. Using specialized sampling equipment, they sought new organisms in the sea and looked for evidence of life in space or on other other planets. Unusual specimens that were collected from remote, inaccessible regions helped test current theories of physiological adaptation and increased understanding of how diverse forms of life evolve. The discoveries in the ocean depths fostered wonder at the variety and complexity of life-forms found on earth. The entry of humans into space raised hopes that science-fiction tales of life on other planets would be superseded by reliable data on the possible existence of alien life-forms. Speculation about extraterrestrial life stimulated fresh evaluation of the specific properties and limitations of life on earth.

31. Marine Biological Laboratory interior, 1888. The MBL at Woods Hole held its first session in 1888. The laboratory room housed aquaria for live specimens and tables for microscope work. (The vials contained histology stains.) Note the presence of large windows and oil lamps for illumination—electrification of laboratories was still in the future. Researchers at the first session came from Bryn Mawr, the Massachusetts Institute of Technology, the University of Michigan, Mount Holyoke, Vassar, and Wellesley.

32. MBL exterior, 1896. These wooden buildings—erected in 1892, 1894, and 1896, respectively—housed laboratories, classrooms, and the library. The MBL quickly became a center for research and education in American biology and hosted scientists from throughout the world at its weekly seminars.

33. Naples Zoological Station, about 1900. The station was founded in 1874 by the German zoologist Anton Dohrn, who called it "a battlefield where all the different zoological armies (systematists, anatomists, physiologists, and embryologists) may meet and fight their common adversaries (error and ignorance)." It drew scientists from many countries and was the model for countless other biological stations.

34. Inside the Naples station, 1908. The operating room for the experimental physiology laboratory was part of a 1903 addition to the station. Fully outfitted research labs equipped with a closed seawater system allowed scientists to use the latest research techniques to study marine organisms. Scientists at the station paid an annual fee to rent research tables.

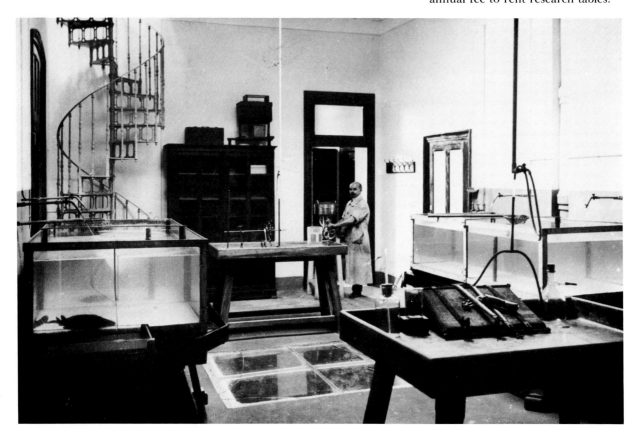

35. Jungle station, Colombia. This temporary site for research on the transmission of yellow fever was set up in 1935. Its equipment included a hand centrifuge and a Primus stove-sterilizer. Urban varieties of the fever transmitted by the mosquito *Aedes aegypti* were more readily controlled than the jungle varieties that were studied here.

36. Collecting mosquitoes. Research on mosquitoes that carry the yellow fever virus, as well as studies of the habits and locale of the disease's monkey and opossum hosts, led to an understanding of the transmission of yellow fever and to the development of a successful vaccine. There are collecting stations at two heights on the same tree in this mid-1930s photograph from the Colombian jungle.

37. Pasteur Institute, Paris. Founded in 1888 for the study and treatment of rabies, the Pasteur Institute developed into a medical research center of international reputation, especially in microbiology. Through their research and teaching at the institute, such scientists as Émile Duclaux, Émile Roux, Charles Chamberland, and Élie Metchnikoff laid the foundations of bacteriology and immunology. Many other Pasteur Institutes were founded both inside and outside France. The photograph was taken about 1904.

39. British National Institute for Medical Research (NIMR), Mill Hill, London. From 1913 the British Medical Research Committee and its successor, the Medical Research Council, supported work in diverse institutions. During World War I these included laboratories at St. Mary's Hospital, the Lister Institute, the London Hospital, and an NIMR facility at Hampstead. The NIMR made fundamental contributions to bacteriology, virology, pharmacology, chemotherapy, and applied physiology. Its directors included the physiologist Henry Hallett Dale, a Nobel laureate. The photograph was taken at the opening of the Mill Hill building in 1950, when the Division of Biophysics alone housed two electron microscopes, electrophoresis apparatus, ultrasonic equipment, electronic counters for radioactive isotopes, a mass spectrometer for stable isotopes, and special freeze-drying and centrifuging equipment.

38. Kaiser Wilhelm Institute for Medical Research, Heidelberg. The Kaiser Wilhelm Society (after World War II, the Max Planck Society) established a series of major research centers in Germany. It opened the Medical Research Institute in 1930 to bring the insights of the basic sciences—physics, chemistry, physiology, and pathology—to bear on the problems of clinical medicine; the buildings included living quarters (*left foreground*) and animal pens (*right*). In the life sciences the society already had institutes for biology, cell physiology, biochemistry, and anthropology.

40. Experimental wheat plots.
England's Rothamsted Experimental Station was established in 1843 by the agriculturist John Lawes and the chemist Joseph Henry Gilbert. Its Broadbalk wheat field, seen here in a 1925 photograph, had been used since 1834 for fertilizer tests.

41. State agriculture station. These tobacco plots are at the Central Crops Research Station at Clayton, North Carolina. Agricultural experiment stations were initially established in the United States in the late nineteenth century to test commercial fertilizers and went on to play an important role in the growth of genetics, bacteriology, biochemistry, and soils sciences.

42. Roscoff Biological Station. The Roscoff station is located at a point on the northeast coast of Brittany particularly well suited to the study of coastal oceanography: several different ocean currents interact, and the tidal range is 10 meters. At the time of this 1987 photograph, major year-round research programs were carried on by 120 full-time staff and 300 visiting scientists in such fields as cellular, developmental, and reproductive biology, neurobiology, marine biotechnology, comparative physiology, and ecology.

43. Collecting in the intertidal. The marine biologist Edward Ricketts, here gathering marine organisms on the California coast, was the model for Doc in John Steinbeck's novel *Cannery Row.*

44. Benthic sampling off Roscoff. The *Pluteus II,* one of the Roscoff station's research vessels, takes samples from the ocean bottom. Biologists collect planktonic and sea bottom organisms with special trawls, tow-nets, dredges, and sampling devices.

45. Tektite underwater habitat. The ocean bottom is not only a site of organisms for study but also a place where the physiological and psychological reactions of humans to a hostile and isolated environment can be investigated. Information obtained in this way helped in the selection of astronaut crews for spaceflight. Experiments with divers living in undersea habitats have included the American Tektite program on the ocean bottom off St. John in the Virgin Islands. Tektite I in 1969 housed four scientist-aquanauts for 60 days. The following year, in Project Tektite II, several diving teams spent time in the habitat, including one group of female aquanauts, here seen leaving the structure.

46. Seatopia. A similar undersea laboratory was this, Japan's first underwater habitat, built in 1971.

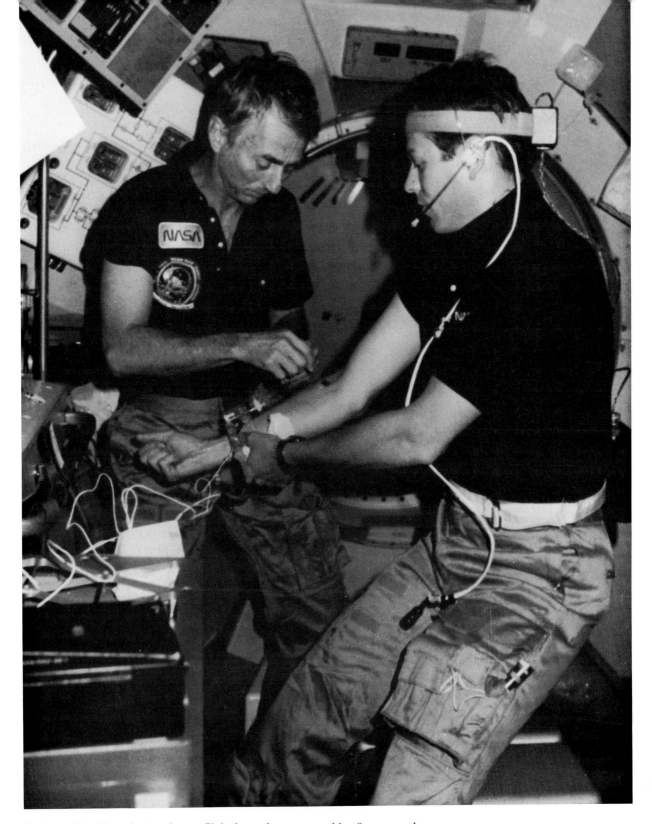

47. Spacelab. The advent of spaceflight brought opportunities for research beyond the earth's atmosphere. Facilities for experiments in many fields were provided in Spacelab, a joint U.S.-European project carried aloft by the U.S. space shuttle. In the scene shown here, from the Spacelab mission of late 1983, one crew member is drawing blood from another for later processing and study on earth.

48. Behring and guinea pig. The guinea pig derives from South America, where it was long a source of food. Introduced into Europe, it was bred as a pet for nearly 300 years before being widely adopted for medical research in the nineteenth century. The German bacteriologist Emil von Behring used guinea pigs as a source of blood serum containing the antitoxin effective against diphtheria. This photograph of Behring with colleagues Wernicke and Frosch was taken in 1891. The year before, Behring had published his first papers on blood serum therapy. For his research on serum therapy, particularly against diphtheria, Behring in 1901 received the first Nobel Prize in physiology or medicine.

4

Organisms

The shift to laboratory research that began in the late nineteenth century markedly altered biologists' needs for experimental animals. Previously, anatomical and physiological studies had routinely made use of stray dogs and cats or farm animals. Medical researchers had also used some smaller animals, such as rabbits and guinea pigs. Around the turn of the century, however, as new specialties like biochemistry and nutrition emerged, scientists began breeding colonies of rapidly reproducing rodents (notably rats and mice) for experimental research. Colonies of opossums and primates were later established for reproductive studies. Obtaining and maintaining animals became a major task of research programs.

In society at large, however, this led to serious questions. Were laboratory animals receiving humane care and treatment? Did the possible benefits to humans—such as a new treatment for disease or new knowledge about how a body system functions—justify causing animals pain and suffering?

From early in the century, medical research in physiology, bacteriology, immunology, nutrition, and biochemistry required a regular supply of healthy experimental animals. Laboratories in these fields established the first full-scale animal colonies. Physiological research was particularly controversial, since physiologists used vivisection, or surgery on live animals, to study respiration, circulation, excretion, the transmission of nerve impulses, muscle responses, and drugs' effects on tissues.

Research that focused on human disease was perhaps less controversial than research on basic physiological processes. But the study of infectious disease did require experimentation. In the late nineteenth century Robert Koch had set forth fundamental rules of procedure for showing that a given microorganism causes a particular disease: the microorganism must be present in every case, it must be obtained in a pure culture, it must produce the disease when inoculated into susceptible animals, and it must be recovered from the experimentally diseased animal. Other medical experimenters working with laboratory animals produced syndromes similar to human diseases by removing a pancreas or thyroid gland. The development of such animal models became even more important in the twentieth century, especially for testing new therapies. As early as the 1890s scientists' arguments for employing animals in research gained support from the production of new biological products like antitoxic serum and organ extracts that had direct therapeutic uses.

All kinds of animals were used. Some of the traditional research subjects have already been noted—rabbits, cats, dogs, guinea pigs, and farm animals, such as horses, pigs, and chickens. Traditional subjects also included frog eggs and tadpoles, and frogs continued to serve as cheap and ready subjects for embryological and developmental studies. Mice proved quite suitable organisms for laboratory research in the emerging sciences of genetics, biochemistry, and nutrition.

Scientists bred special laboratory strains of mice for experimental purposes. The laboratory mouse and rat were convenient and relatively inexpensive and noncontroversial experimental animals. In genetics, however, the tiny fruit fly *Drosophila* largely supplanted the rat and mouse. It was very easy to maintain *Drosophila* in milk bottles, and huge collections of the insects crowded the early geneticists' "fly rooms."

Cockroaches, too, found their way into special laboratory colonies for research on insect hormones and, later, animal behavior. Neurophysiological studies employed squid. Throughout the century cell and developmental research made use of simpler marine organisms, like sea urchins, whose rapid transformations from egg to multicellular embryo to adult could be readily viewed under the optical microscope.

Aristotle in the fourth century B.C. had studied development in bird eggs. Now, scientists used adult birds such as chickens, canaries, pigeons, and ducks to investigate animal behavior, as well as such diverse problems as the transmission of infectious disease (a notable early example being Ronald Ross's Nobel-winning work on avian malaria in the late nineteenth century), the role of vitamins and minerals in diet, and the effects of hormones on growth and development. Chickens, like several other agriculturally important animals and plants, proved of especial interest to agricultural scientists and geneticists, as well as to medical researchers.

In the twentieth century biologists finally succeeded in tracing the life cycle, and determining the detailed structure, of microorganisms such as bacteria and viruses. This achievement was partly due to the astonishingly high magnification provided by the electron microscope—but partly as well to the invention of specialized techniques for culturing bacteria and cells. Research on the life cycles, genetics, and biochemical processes of these simplest forms of life revealed complex interrelationships between microorganisms and human, plant, and animal diseases. Microorganisms also provided simple examples of basic biochemical processes.

Unanticipated and far-reaching opportunities ensued from the development early in the century of special culture techniques for isolated mammalian organs and tissues. Cell culture techniques—which involve removing cells from their normal tissue environment and providing them sustenance so they grow and reproduce—allowed scientists to study the inherent properties of specific cell lines or of clones formed from single cells. After mid-century, biologists regenerated whole organisms from single cells and studied in detail the interrelationship between cell form and the external environment in which cells grow. Perhaps the most famous human cell line was HeLa, named after the woman (a cancer patient in Baltimore) from whom the cells were originally taken in 1951. These tumor cells, cultured in the laboratory, played an important role in medical research in the following decades.

During the century biologists increasingly employed experimental animals to test newly developed pharmaceutical products—vaccines and drugs, including antibiotics—to help ensure that the products would be effective and safe in humans. Researchers continued to make extensive use of larger mammals, notably dogs, cats, and monkeys, in physiological studies of brain and nervous-system function and for special missions into environments like outer space. The first traveler beyond the earth's atmosphere was not a human being but a dog, Laika, carried aloft by a Soviet spacecraft in 1957.

Beginning about the 1970s debate sharpened over the role of monkeys, chimpanzees, and other primates in medical and biological research. Many people saw primates, like dogs and cats, as having a special relationship to human beings, and the old antivivisection controversy reawakened with new vehemence. Questions were raised again about the necessity and value of certain animal experiments in biomedical research, and critics protested the routine use of animal tests in the cosmetic and chemical industries. As a result, the scientific establishment stiffened criteria for the maintenance, care, and use of laboratory animals in research.

49. Slye and mice. Between 1913 and 1941 the American pathologist Maud Slye bred thousands of mice to study the inheritance of cancer. Her view of cancer as a single recessive trait was challenged by the experimental biologist Clarence Little, who, in contrast to Slye, worked with inbred strains.

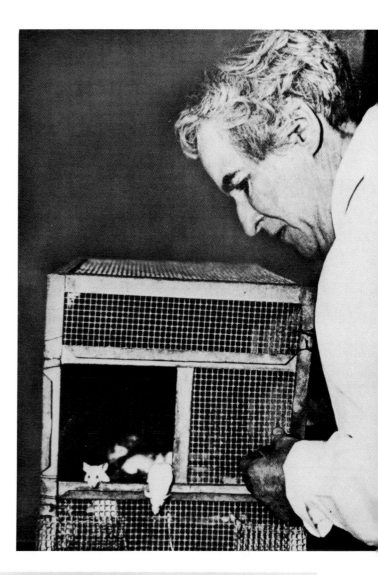

50. Inbred mouse and pups. These mice are from the so-called DBA strain, the oldest inbred strain on earth. Developed by Little in 1909, the DBA strain has been used to establish mouse colonies throughout the world. Inbred experimental animals provide researchers with subjects that show little or no genetic variation, making it possible to avoid the effects of unknown or variable mixtures of genes on experimental studies. Little saw that "just as the chemist needed pure chemicals, the biologist needed genetically pure strains if he was going to succeed in analyzing the causes of various complex medical traits, especially cancer." DBA females tend to develop breast tumors, and in 1933, Little and his colleagues suggested that a viral agent—later called mouse mammary tumor virus—caused the tumors. Both genetic and nongenetic factors are now thought to be involved in the development of these tumors in mice.

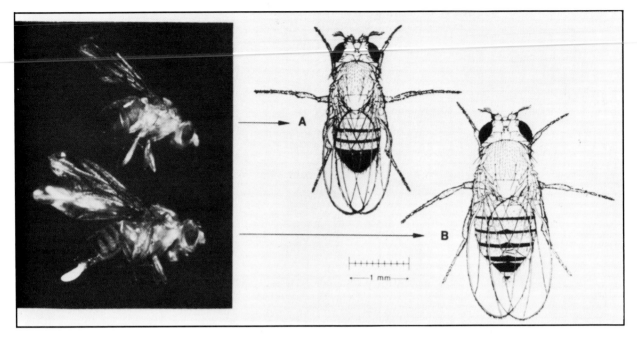

51. *Drosophila melanogaster.* Because they are easy to raise and study and take only about ten days to double in number, these fruit flies, along with others in the genus *Drosophila*, have been used extensively in researchers' experiments on heredity and evolution. Shown here is the normal, wild-type *Drosophila melanogaster,* with the male indicated by **A** and the female by **B**. The wild-type has a gray body and dull-red eyes. *Drosophila* larvae hatch from the egg in one day (adults emerge after another four days), and females lay several thousand eggs during their lifetime.

52. *Drosophila* laboratory. So easy are fruit flies to raise, scientists have often grown them in milk bottles, using bananas as food. Hence the multitude of bottles in this room at the University of Texas at Austin, which emerged as a major center for *Drosophila* research in the period after World War I. All four men here made substantial contributions to genetics: T. S. Painter (seated at rear), W. S. Stone (standing behind trays of bottles), C. P. Oliver (seated in front), and H. J. Muller (examining flies with a jeweler's loupe).

53. Ross, with bird cages. The British bacteriologist Ronald Ross, working in India, verified the theory that the malaria parasite is carried by mosquitoes. In the final stage of his research he determined the parasite's life cycle using caged birds, ultimately tracing the development of the parasite in the mosquito. In 1902 he received the Nobel Prize for this work. The life cycle of the human malaria parasite was elucidated only after Ross explained the avian form of the disease. In the photograph, taken in Calcutta in 1898, the year he completed his proof of the mosquito theory, Ross is accompanied by his wife and assistants.

54. Effects of castration. Compare the cockerel, or young male, of a White Leghorn chicken at right to the capon, or castrated male, at far right. Castration has altered the comb, wattles, and ear lobes. Observation of the changes wrought by castration in birds and other animals has traditionally been an important technique for analyzing the function of the sex glands and their hormones. Male hormones were first isolated and synthesized in the 1930s.

55. Stanley examining tobacco plants. Are viruses alive? Some thought the answer was no, after the American biochemist Wendell Stanley used chemical techniques to crystallize the tobacco mosaic virus in 1935. The virus had not been amenable to isolation by ordinary bacteriological procedures. Here, Stanley (*right*), with his German-born colleague Heinz Fraenkel-Conrat, is studying some infected tobacco plants; the plants in the foreground are uninfected.

56. Varieties of viruses. The invention of the electron microscope in the 1930s eventually made it possible to actually see viruses. These 1940s electron micrographs show, to the same scale, the following viruses: (1) vaccinia, (2) influenza (Lee strain), (3) tobacco mosaic, (4) potato-X (latent) mosaic, (5) bacteriophage T2, (6) Shope papilloma, (7) Southern bean mosaic, and (8) tomato bushy stunt. Subsequent study, especially of the bacteriophage, led to the rise of molecular biology and the discovery of the genetic code.

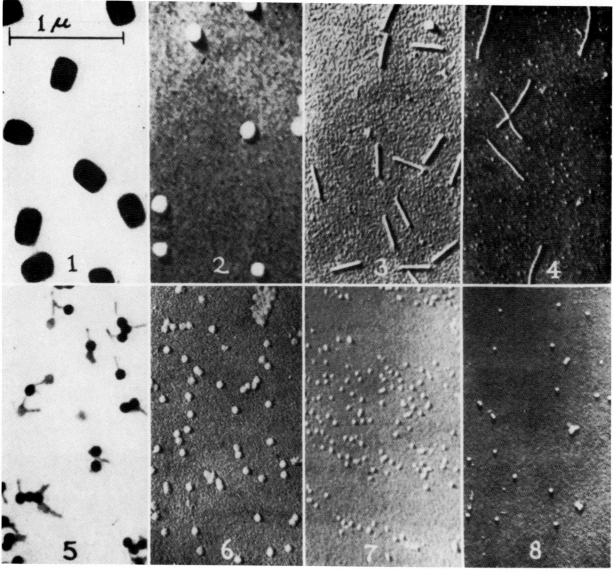

57. Founder of the HeLa line. In 1951 a 31-year-old black American woman known as Henrietta Lacks or Helen Lane (the hospital record was destroyed) died in Baltimore of cervical cancer. Cells from her tumor were preserved and, multiplying, became the first continuously cultured strain of cancer cells. Named HeLa, they found application in laboratories around the world in research on cell growth, protein synthesis, and the biology of viruses.

58. Carrot plant, cloned. A pioneer in the cultivation of plant tissues was the U.S. physiologist Philip R. White, who in 1934 showed that excised root tips of the tomato plant could be made to grow. By about 1960 another American, Frederick Steward, succeeded in regenerating entire carrot plants from single root cells.

59. HeLa cells. The HeLa strain develops extraordinarily well in laboratory cultures. Typical of cancer cells, HeLa cells have large or multiple nuclei and grow densely. Although techniques for culturing tissue were developed early in the century, cell culture methods for obtaining clones or colonies from single somatic (that is, nonreproductive) cells were not perfected until the 1950s.

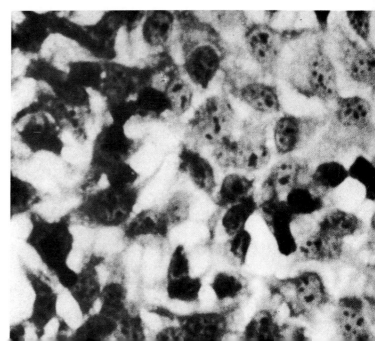

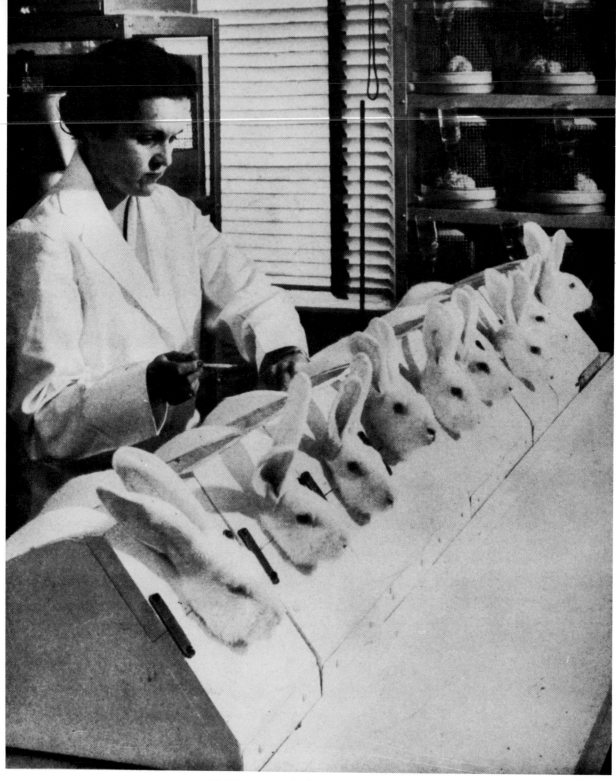

60. Pharmaceutical quality control, 1940s. Here we see a test for penicillin impurities that could cause fever; penicillin that made rabbit temperatures rise was discarded. Routine use of animal tests in medical research and in the pharmaceutical and cosmetic industries led to a resurgence of animal rights groups in the latter part of the century. (Antivivisectionist groups had protested the use of higher animals in research since the mid-1800s.) Such concerns led to increased regulation of animal housing and care, and in some countries it was required that scientists making use of animals in research be licensed.

61. Laika: first space traveler. Farm animals, as well as domestic animals like dogs and cats, have long been used as subjects in physiological research. They helped space medicine develop as a distinct field of research after World War II, as standard physiological monitoring techniques were adapted for the specific conditions of spaceflight. Laika was carried aloft in a Soviet spacecraft in 1957.

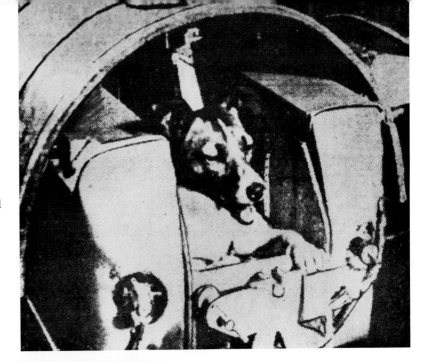

62. Sam. The century saw primates become common subjects for biomedical research, especially for physiological and psychological studies requiring subjects closely related to humans. Sam, a 7-pound rhesus monkey, traveled 200 miles into space in December 1959. One of the purposes of the flight was to test the type of capsule to be used by astronauts in the U.S. Mercury space program. Sam rode in a 100-pound biopack container strapped in the capsule's flight couch. (Attached to the container were other biological specimens such as barley seeds, molds, and flour beetle eggs.) Biomedical data on spaceflight were an object of the mission, and Sam underwent tests after his return to earth.

63. The new observational technology. Methods of observing the body multiplied rapidly during the century, as researchers perfected instruments and tests for peering into the living organism. Patients often felt invaded by the diagnostic eye of medicine, whose techniques were increasingly dependent on the use of precision apparatus. New industries arose to equip laboratories, hospitals, and clinics with these specialized observational tools.

5

Methods of Observation

The traditional basic research technique of biological scientists is simple, naked-eye observation of living things. But twentieth-century researchers came to employ diverse observational methods. Many scientists, combining more and more skillfully techniques derived largely from chemistry and physics, turned to observation of internal and subcellular events.

At the beginning of the century the principal aid to observation was the optical microscope. It had come into use in the 1600s but did not emerge as a powerful research tool until the nineteenth century, when scientists eliminated chromatic and spherical aberrations and devised special stains for specimens. Also, in the mid-nineteenth century physicians had developed additional viewing devices, or scopes, for observing the body externally and internally—such as the ophthalmoscope for the eye and the laryngoscope for the vocal cords.

The optical microscope eventually had to share its preeminence with the even more powerful electron microscope. Invented in the early 1930s, the electron microscope became commercially available by late in the decade. Its resolving power surpassed that of the optical microscope, revealing structures formerly inaccessible to researchers and eventually exposing cellular events to observation and detailed analysis. After researchers developed new specimen-sectioning and contrast techniques, they could determine the fine structure of cells and study the biology of bacteria and viruses. In time,

the electron microscope exposed macromolecules like DNA to view.

Viruses were first studied with the electron microscope in 1939. Metal-shadow-casting and freeze-drying techniques, developed in the 1940s and 1950s, showed these organisms to be structurally diverse. Viruses presented enormous theoretical interest because many scientists thought them to be relatively simple biological entities.

In studies of cells of higher organisms, the optical microscope and electron microscope complemented one another. The nature of the phenomenon under investigation determined which instrument was more useful. Histological research on tissues (that is, groups of cells) with the optical microscope was supplemented with detailed electron-microscope studies of individual types of cells exhibiting special properties or responsive to specific drugs or hormones.

The internal structure of the cell, revealed as immensely detailed under the electron microscope, became of foremost interest after mid-century. Biologists identified and named new subcellular particles and debated their functions. Researchers increasingly sought to ascertain the location within the cell of specific biochemical events. They succeeded in localizing biochemical events within particular "organelles," in part by applying other analytical techniques to the structural problems studied by microscopists.

Another crucial imaging technology was the X ray, discovered and first utilized medically in the

1890s. Like the microscope, X rays revealed hidden internal structures and exposed internal processes to analysis. Fluorescent tracking agents were an additional boon. With these, physicians could follow the movement of substances through the digestive tract, investigating normal physiological processes and diagnosing pathological alterations in them. In a similar way, radioactivity later became an important method for imaging biological events. By exposing photographic plates to biological structures that had incorporated radioactive materials, researchers could reveal the presence of specific sites of chemical synthesis within the cell. This made it possible to observe and track the process of biochemical change without the disruptive effects of chemical analytical techniques. Radioactive tracers came into extensive use after World War II. These enabled biologists to study how molecules are localized, utilized, and transformed in processes like photosynthesis, both in individual cells and within the organism as a whole.

The electronic computer revolutionized imaging technology. The computer could modify or even entirely construct images. Some structures not ordinarily revealed by standard microscopic or X-ray techniques could now be seen without intruding into the body. For example, computerized tomography showed soft-tissue tumors not visible in ordinary X-ray views. Such techniques proved especially valuable in medicine, since in many cases they eliminated the need for invasive measures in making a diagnosis.

Of course, one of the most obvious legacies bequeathed to the twentieth century by its predecessor was photography. By combining magnification with photography biologists could evaluate images produced under different physiological conditions. They could study, compare, and contrast images with one another. They used time-lapse photography, moving-picture films, and, later, video tape to reveal and preserve for analysis transient biological events such as cell division.

64. First electron microscope. More precisely, this is the first commercially produced electron microscope, made by the Siemens and Halske Company in Germany in 1939. The electron microscope, developed principally by the German physicist Ernst Ruska earlier in the decade, was one of the most important inventions of the twentieth century, opening up to observation a world of detail previously inaccessible to the human eye. With the electron microscope, the structure of viruses and bacteria was revealed for the first time, as was the fine structure of animal and plant cells. Commercial electron microscopes that were reliable, compact, and easy to use came on the market after World War II.

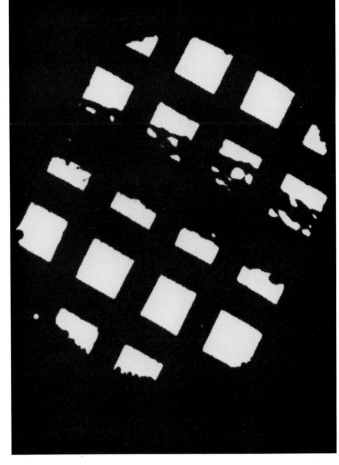

65. Early electron micrograph. One of the first electron-microscope photographs of a biological specimen was this picture of the root of a bird's-nest orchid. The Hungarian-born physicist L. Marton made it in 1934, using an instrument he built himself. The resolving power of the earliest electron microscopes could not equal that of the best optical microscopes, but by the 1940s electron microscopes were producing images of viruses and macromolecular crystals.

66. Optical microscopy: preparing to observe. While the electron microscope was perfected and refined, the optical microscope remained a powerful observational tool and underwent refinements of its own. The microscopist in this 1984 scene is getting ready to photograph radioactive waste samples. He selects a suitable objective lens from the variety available.

67. Electron microscopy: inserting the specimen. The electron microscopist in this mid-century photograph is the American biophysicist Robley Williams. He codeveloped the technique known as metal shadow casting, used to increase the contrast of a specimen and create a three-dimensional effect. Williams, who also made contributions to the study of viruses, was one of several notable scientists in this century who moved into biology after training and working in physics.

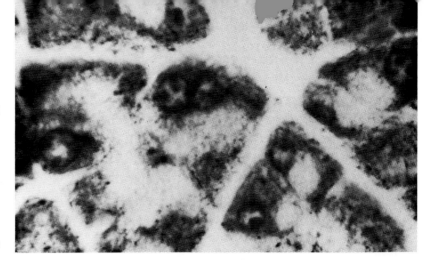

68. Viewing the pancreas through optical and electron microscopes. At right, above, is a "high-power" optical micrograph of exocrine, or secretory, cells in a mammalian pancreas. One can make out the nucleus of each cell, but it is hard to discern any detail other than granules in the cytoplasm—the part of the cell outside the nucleus. Below is an electron micrograph of the same kind of cell. Here the cytoplasm shows much structural detail, including membrane systems, the tiny energy-supplying bodies called mitochondria, and lysosomes—bodies containing digestive enzymes. In the inset, at the upper right of this micrograph, a much greater magnification reveals the ribosomes, where proteins are synthesized. Before the electron microscope scientists did not suspect that such fine internal structure existed.

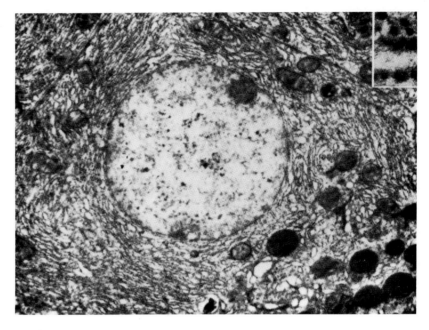

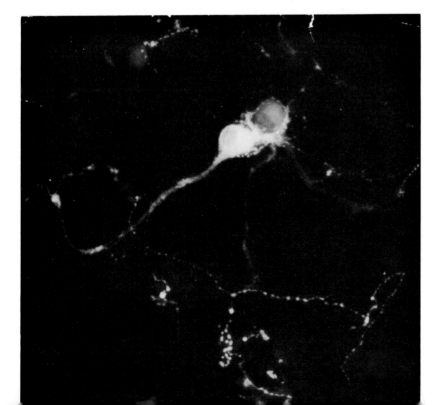

69. Fluorescence microscopy. In this refinement of conventional optical microscopy, the specimen is specially treated and then subjected to blue or ultraviolet light. It fluoresces, revealing internal structures or the concentration of particular chemicals in certain areas. The fluorescent photomicrograph presented here shows mouse spinal column cells that contain enkephalin, a kind of natural painkiller produced by the body.

51

Ея Имп. Величества Государыни Императрицы
Александры Феодоровны
лѣвая рука.

71. *The Fluoroscope.* For the practice of medicine, X rays were a boon. The fluoroscope combined X rays with a fluorescent screen so that the interior of the body could be immediately seen by the doctor. In this 1926 etching by the American artist John French Sloan, the patient is Sloan himself. The physicians are observing the movement of fluid down the digestive tract, through the stomach and intestines. Much of Sloan's work concerned the everyday experience of individuals in the city.

72. First biological autoradiograph. X rays were not the only discovery of the late nineteenth century with implications for imaging. Radioactivity was used in making autoradiographs, or radioautographs. When placed on a photographic emulsion, an object containing a radioactive material produced an image. By introducing a radioactive substance into a living animal or plant, biologists could study how that substance was subsequently distributed in the organism's tissues. This first published (1908) autoradiograph of a biological specimen was made by placing a frog into a container with radium and then on a photographic plate. The radium, taken up by the skin of the frog, produced an image of the entire animal.

◀ 70. X ray: the czarina's hand. The discovery of X rays at the end of the nineteenth century by the German physicist Wilhelm Konrad Röntgen (Roentgen) made a new observational tool available to biologists, allowing noninvasive views of the interior of the body. This early X ray, taken in 1898, shows clearly the bones as well as the jewelry of the left hand of Empress Aleksandra Feodorovna of Russia. The potential danger of exposure to X rays was not immediately recognized. For the public, X rays were a curiosity, meriting, at least in the case of a czarina, an artistic frame.

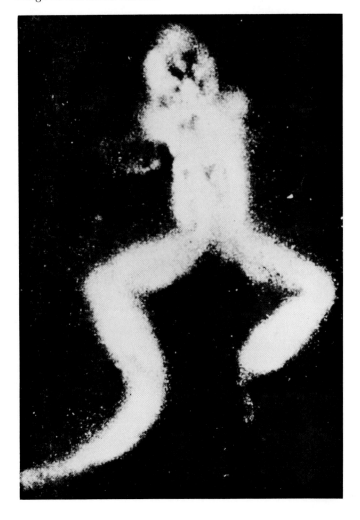

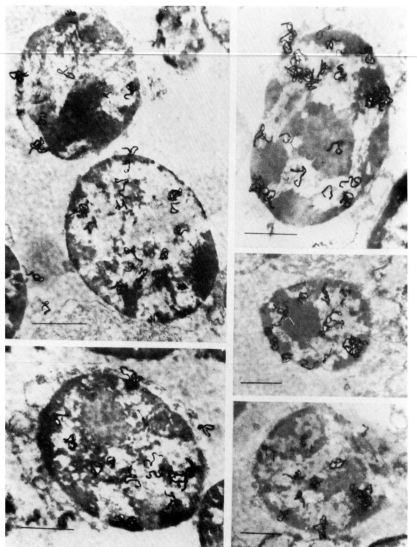

73. Observing biochemical events in the nucleus. These autoradiographs published in 1964 portray the formation of ribonucleic acid (RNA) molecules, which play a key role in the transmission of genetic information in the cell nucleus. Researchers treated the nuclei of calf thymus cells with a radioactive form of a substance utilized in the synthesis of RNA. The photographic silver grains show where the substance was taken up and incorporated into RNA. We are able to observe the results of this process of synthesis, called transcription, thanks to the electron microscope, which provides the necessary magnification for seeing the nucleus's fine structure.

74. Autoradiograph of DNA. Here the DNA, or basic genetic material, in the loop chromosome of the bacterium *Escherichia coli* was "labeled" with a radioactive substance and separated from the rest of the bacterium by the use of a purified digestive enzyme. Such procedures helped illuminate the new perspective that was provided after mid-century by molecular biology.

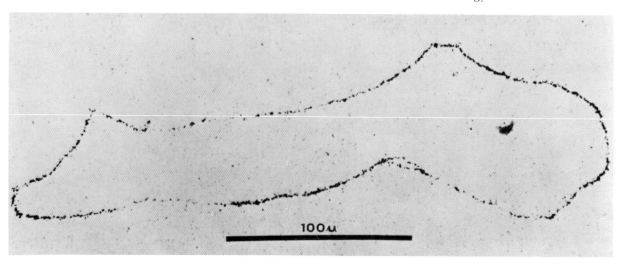

100µ

75. First CT scan. The rapid development of electronic computers after World War II made possible new imaging techniques, which proved of value in medical diagnosis. Computerized tomography uses X rays to produce a cross-sectional image composed of numerous tiny "dots" that reproduce the structure of soft tissues, showing internal organs and abnormal growths. These CT head scans were the first on a human patient. The patient had a tumor in the left frontal lobe of the brain, revealed as a dark area in the scan.

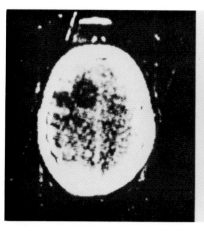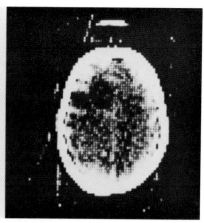

76. Operating microscope. Enhanced by new technologies, the microscope developed into a valuable aid for performing delicate surgical procedures. The binocular coobservational microscope used by these surgeons (who are working on the larynx of a horse) magnifies the surgical field—the nerve of interest is only 2 millimeters in diameter. The image is televised at the rear and videotaped for later study.

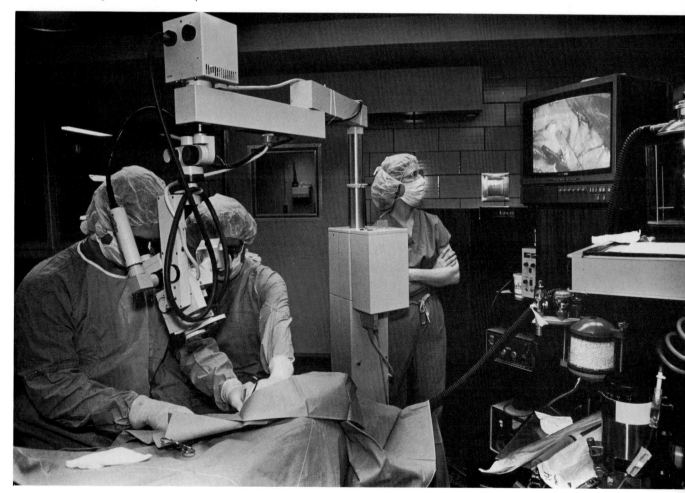

77. Mathematics and biology. The British mathematician Karl Pearson, seen here in 1910 with his calculator, has been called the father of the science of statistics. In the last decade of the nineteenth century he undertook to develop statistical methods for the study of genetics and heredity. A professor at University College in London, Pearson established a "biometric" laboratory where a number of noted statisticians of the early twentieth century trained. He also edited the influential journal *Biometrika*, which he helped found in 1901. His contributions to statistics, including the chi-square "goodness-of-fit" test for the significance of experimental results, found broad application in the biological, social, and physical sciences alike.

6

Modes of Analysis

The twentieth century saw biologists increasingly supplement their traditional research method—observation—with new analytical techniques. They often borrowed these techniques from physics and chemistry, although in the case of statistical analysis, the flow of ideas might be said to be in the other direction. The mathematician Karl Pearson's application of statistics to the biological and social sciences yielded methods of importance to the physical sciences as well. Be that as it may, routine measurement, statistical analysis, graphic presentation of data—all common, indeed identifying, features of the modern scientific method—entered the life sciences in the latter decades of the nineteenth century. At that time new recording instruments and standardized chemical techniques began to transform biology from a primarily descriptive endeavor into a rigorous experimental science. As research and teaching laboratories were established and investigators adopted and refined these analytical tools, biological disciplines grew more numerous and more specialized. A variety of new specialties arising from bacteriology, biochemistry, and related fields trained students for jobs dependent on newly developed analytical laboratory skills.

In the nineteenth century, physiologists were the first life scientists to use graphic ("kymographic") recording instruments, which they adapted from physics and meteorology; the devices then spread to other areas of biology. With these instruments researchers could record very

rapid transient events. Kymographic recording techniques profoundly affected experimental design in subjects as diverse as botany and animal behavior. Although originally constructed to record mechanical events—movements or pressure variations (as of blood)—the instruments were eventually adapted to monitor minute electrical events in nerves, muscles, and the heart. Early in the twentieth century biologists correlated the resulting data on internal physiological events with newly acquired information on the internal chemical environment and the biochemical activity of tissues. They became intensely aware of the extraordinarily complex and coordinated processes that occur in living organisms, and they sought to understand these regulatory mechanisms.

Researchers studying processes within organs still made use of the traditional vivisectional methods—that is, the surgical exposure of living tissue, either in order to excite a nerve or muscle or to take samples of blood or other fluids. But in the 1920s to 1940s these surgical methods combined with new microdissection, microinjection, and microsampling techniques to provide increasingly precise information about the location and extent of chemical transformations within certain organs. Scientists began to understand photosynthesis, respiration, and cell metabolism in terms of subcellular biochemical events dependent on catalysis by enzymes.

At first, researchers concentrated on isolating specific substances produced by tissues or cells—

substances like enzymes, vitamins, hormones, and antibodies. But this gave way in the thirties and forties to detailed analysis of biosynthetic pathways, the precise sequences of reactions by which substances are transformed in cells. Study of large molecules, subcellular particles, and organelles—isolated through the ultracentrifugation techniques invented by the Swedish chemist The Svedberg—helped scientists locate and analyze proteins, enzymes, and metabolic products within cells. Beginning in the 1940s tracer atoms, usually radioisotopes, made it possible for researchers to track the fate of discrete molecules within specific tissues. The powerful separation techniques of chromatography and electrophoresis greatly aided such studies. The former was introduced by the Russian Mikhail Tsvet early in the century, and the latter by the Swede Arne Tiselius in the 1930s. These and similar physical methods for isolating sizable biological molecules facilitated analysis of the degradation products of larger molecules and the identification of intermediate molecules produced by rapidly occurring chemical processes in organisms.

The powerful new chemical and physical techniques came to be applied routinely in identifying the products of biological activity. But their usefulness rested in part on the ability of the researcher to isolate experimentally a specific biological process for analysis and study. This was possible thanks to the refinement of manipulative isolation techniques derived from surgery and to the development of specialized methods for culturing organs, tissues, and cells.

The new organ, tissue, and cell culture methods gradually replaced vivisectional techniques. But the culture methods did not just allow the researcher to isolate and study specific processes. They were also adapted for the commercial production of many biological molecules—products as diverse as vaccines, hormones, vitamins, enzymes, and antibiotics. Beginning in the 1970s such substances were increasingly "manufactured" from artificial sources—that is, from non-natural cells or organisms created by transplanting the cell nucleus or splicing genes. In other words, researchers transferred subcellular organelles or molecular fragments from one living unit to another to create organisms with unique metabolic or synthesizing capabilities.

Thus, the century saw life scientists apply specialized analytical techniques used in chemistry, physics, anatomy, and physiology to the study of a wide range of biological problems, creating in the process a new, experimentally oriented biology. Research focused more and more on functional problems, explaining how organisms worked, rather than on descriptive and taxonomic questions per se. In the process, the systematic approach that had been characteristic of much pre–twentieth-century zoology and botany slipped into the background.

78. Measuring plant growth. In this 1913 Smith College plant physiology class, taught by William F. Ganong, the students are using auxographs, or recording auxanometers, to measure plant growth. Auxographs were an early instance of botanical studies' adaptation of recording instrumentation from animal physiology. Ganong wrote an experimentally oriented botany text.

79. Radio-monitoring incubation temperatures. While a scientist places a radio transmitter in an egg under a nesting antarctic Adélie penguin, an electronics technician adjusts the aerial for signal strength. Routine monitoring of biological events developed from the nineteenth century's interest in quantification and the recognition that variables change over time.

80. Quantifying crab activity. In a mid-century tableau from an American aquatic biology laboratory, the graphic recorder registers the activities of the crab in the box. When the crab moves, a pen marks the event on the surface of the continuously moving paper at the left. With this graphic record the researcher can, at leisure, "observe" and measure the rate and the intensity of activity.

81. Sampling in the field. In a mid-century scene, physiologist Roy Forster, noted for his work on the vertebrate kidney, takes blood from a goosefish aboard a fishing boat. The development of techniques for sampling and chemical analysis of minute quantities of fluids made it possible for twentieth-century researchers to study chemical processes as they occur in living organisms.

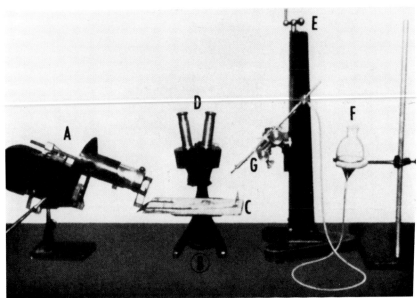

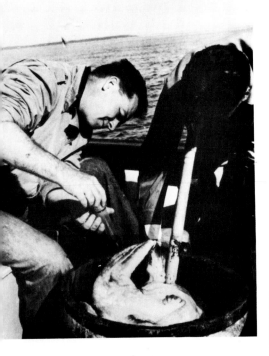

82. Micropuncture sampling. In the 1920s the mechanism of filtration and reabsorption in the kidney was still largely unknown, and the American pharmacologist Alfred Newton Richards and his colleague Joseph Wearn devised this ingenious apparatus for penetrating the tiny glomerular capsule in a frog's kidney to obtain samples of fluid for analysis. With the frog under the microscope, a stand held an extremely thin hollow needle (**G**) that would penetrate the capsule. The needle could be guided mechanically, and its position observed. The adjustable bulb (**F**) contained mercury used to block sections of tubule in the kidney so that fluid in specific segments could be withdrawn for analysis.

83. Drawing a sample in photosynthesis research. By incubating cell parts with radioactive chemicals, scientists can follow the course of chemical reactions that the cell parts initiate: they trace radioactive, or "labeled," molecules along the pathways by which new substances are synthesized. In the photograph, samples of media containing chloroplasts are drawn from the flasks at certain intervals in order to determine how specific molecules are transformed during photosynthesis. The use of such tracers allows analysis of even tiny amounts of fluid.

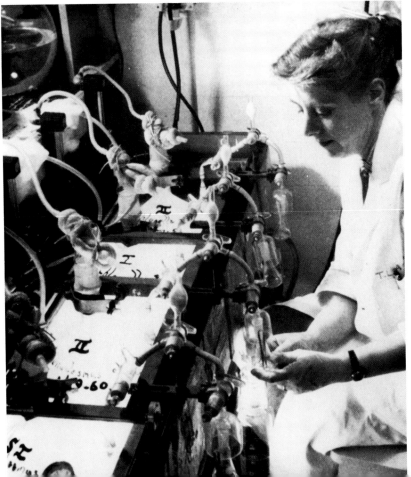

84. Isolation and analysis of biological products: Hopkins and tryptophan. In 1901 the British biochemist Frederick Gowland Hopkins (*right*), together with Sidney W. Cole (*left*), isolated the amino acid tryptophan. In the photograph Hopkins holds the first specimen obtained. A few years later he showed that certain animals cannot make tryptophan and other "essential" amino acids; these thus must be obtained from the diet. Regarded as the father of British biochemistry, Hopkins shared the 1929 Nobel Prize in physiology or medicine for work on "accessory food factors," or what are now called vitamins. The isolation and chemical identification of specific biological molecules became a major focus for biochemical research in the first half of the century.

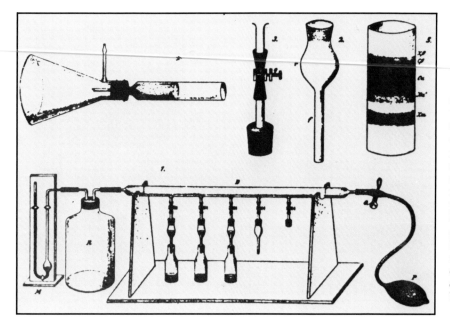

86. Paper chromatogram of insulin components. Paper chromatography, developed in 1944, allows more complete separation of mixtures than Tsvet's method because it is two-dimensional. For this chromatogram hydrolyzed, or digested, insulin was dissolved in a solvent and placed at the upper left-hand corner. The filter paper was put in an air-tight chamber containing solvent, which moved different components different distances along one axis. The procedure was then repeated using another solvent at right angles to the first run-through. The resultant chromatogram provides a fingerprint of the major subunits making up the insulin molecule.

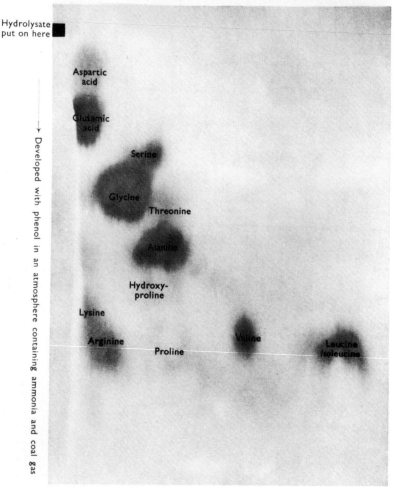

Developed with collidine in an atmosphere containing a trace of diethylamine

Hydrolysate put on here

Developed with phenol in an atmosphere containing ammonia and coal gas

Aspartic acid

Glutamic acid

Serine

Glycine

Threonine

Alanine

Hydroxy-proline

Lysine

Arginine

Proline

Valine

Leucine Isoleucine

87. Electrophoresis of human blood serum. Introduced as a reliable analytical technique in 1937 by the Swedish chemist Arne Tiselius, electrophoresis uses an electric field to move charged particles dispersed in a liquid. Particles of different sizes or charges move at different rates and thus are separated. Electrophoresis rapidly became important in the development of immunology, biochemistry, microbiology, and physical chemistry and was applied to the study of blood sera, antibodies, viruses, enzymes, and proteins. Commercial electrophoresis units became available after 1945. The scan shown here has peaks for (*left to right*) the blood proteins albumin, alpha-one globulin, alpha-two globulin, beta globulin, and gamma globulin.

88. Automatic amino acid analyzer. By the 1960s, apparatus like this was used for analyzing mixtures of amino acids. The vertical tubes in the center are "ion-exchange" columns. At lower left is the recorder, which generates a graph with peaks for the various component amino acids. Pumps drive salt solutions containing the mixture through the ion-exchange column. A coloring reagent is delivered at the base of the column so that the presence of an ordinarily colorless amino acid can be detected. Color intensity is measured by a photometer connected to the recorder.

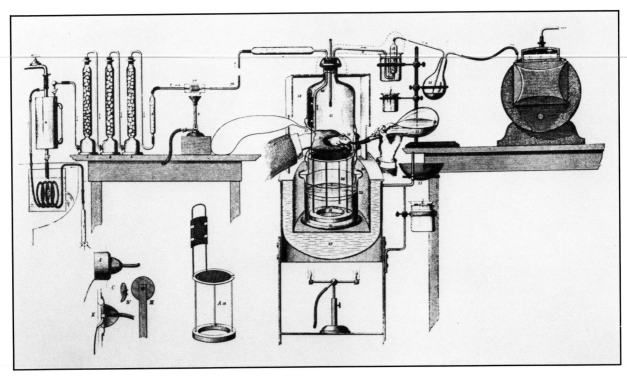

89. First germ-free isolator. George H. F. Nuttall and H. Thierfelder built this apparatus in Germany in 1895, largely to test Pasteur's belief that microorganisms were essential to animal life. Newborn guinea pigs that had been delivered under sterile conditions were placed in the sterilized glass chamber. A rubber glove attached to the wall permitted manipulation of the animals, who were kept alive until the supply of sterile food ran out. No microorganisms were then found in the guinea pigs.

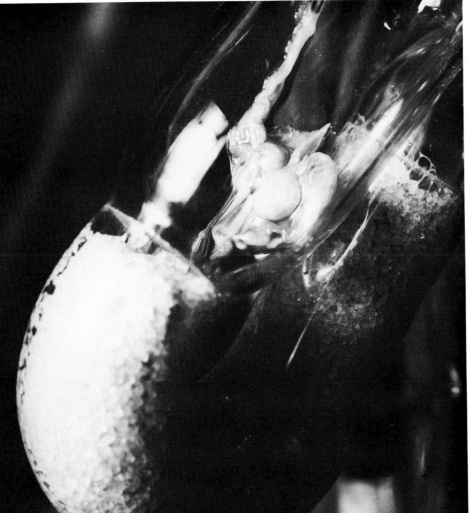

90. Carrel's tissue culture apparatus. The French-born American surgeon and experimental biologist Alexis Carrel placed muscle from a chick heart in the apparatus in 1912. Supplying nutrients and removing waste products, he claimed to be able to keep the cells alive for several decades. While the success of the experiment has been debated (overzealous assistants may have replenished the cells), Carrel's work stimulated a long tradition of research in tissue culture and transplantation that eventually yielded remarkable advances.

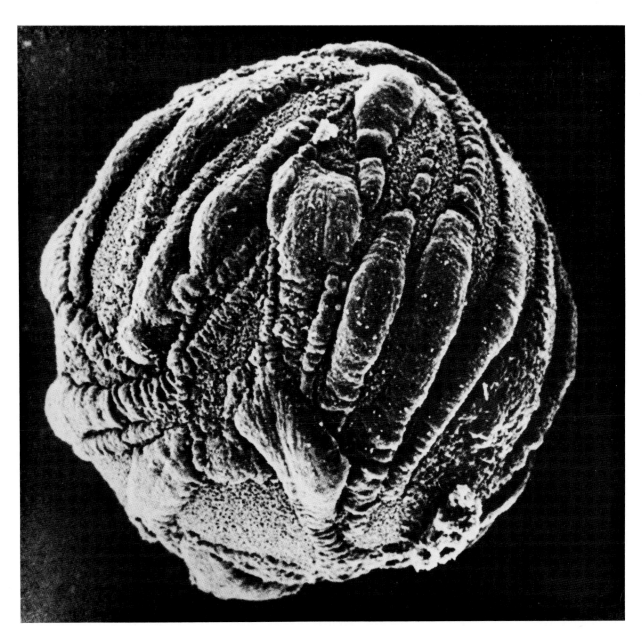

91. Tissue culture for making interferon. This ball is actually a tiny bead made of a synthetic polymer (dextran), seen magnified in a scanning electron micrograph. In a technique in use around 1980, such beads were put into cell cultures to provide a surface on which the cultured cells could grow. In this case, the cells were human foreskin cells, used to make interferons, a class of defensive substances produced by animal cells in response to virus infection. The cells were first grown in a nutrient medium. With the addition of a special interferon-inducing chemical, the cells were transferred to another medium for interferon production. The medium was concentrated, dialyzed, and freeze-dried in order to recover the interferon.

Part Three

TRANSFORMATION
OF THE
NATURAL HISTORY
TRADITION

92. Walcott's quarry. Called by some the single most important fossil find, this site is near Burgess Pass, not far from Field, British Columbia, in western Canada. The American paleontologist Charles Walcott discovered it by accident in 1909. About 35,000 fossils of soft-bodied invertebrates have been found in the bed of rock known as the Burgess shale. They date from the middle of the Cambrian period, which lasted from about 570 million to about 500 million years ago. Many of these soft-bodied fauna were unknown elsewhere. Indeed, soft-bodied creatures are fossilized relatively rarely. The Burgess shale fossils not only reveal the variety of Cambrian life but are remarkable for their fine structural detail, often including internal structure.

7

*P*aleontology

Many of the central theoretical problems of twentieth-century biology were framed by the theory of evolution. They grew out of Darwin's interpretations of discoveries in geology, zoology, and botany. Paleontology spans these fields. It uses fossil remains to study the plants and animals of past geologic eras, focusing specifically on the nature, origin, and occurrence of individual types or species. Employing geological data to evaluate and date specimens, it relies on biological principles to reconstruct fossils' anatomy and genealogical relationships.

The fossil record shows that organisms change in structure over time. Evolutionary theory proposed a mechanism that might account for these changes. Darwin called the mechanism natural selection, suggesting it was a process in nature analogous to the artificial selection practiced by dog breeders and pigeon fanciers. Biologists have challenged, tested, and reinterpreted Darwin's hypothesis, always taking fossils for their initial data.

Twentieth-century paleontologists have frequently worked in museums, like the American Museum of Natural History or the British Museum (Natural History). These institutions house large collections of bones that can be compared with one another for research purposes or reassembled for public display. Skeletons reconstructed by skilled preparators are exhibited in huge halls that present current understanding of presumed evolutionary relationships.

But museum collections depend on field work for their sustenance. Since the nineteenth century, paleontologists—armed with picks and hammers, a pad of paper and pencils—have explored remote areas of the world in search of prehistoric creatures. New finds might suddenly alter the interpretation of older data, and paleontologists do in fact constantly revise evolutionary trees. They gained insight into the evolution of the horse, for example, by correlating the appearance of equine bones within different geologic strata and mapping this evidence over geologic time.

Mammoths and other striking large vertebrate animals were popular objects of study in the nineteenth century. But the twentieth century saw many paleontologists shift their attention elsewhere, to invertebrate forms and, much later, microscopic unicellular fossils. Researchers also focused increasingly on the remote past. The discovery of fossilized unicellular organisms fueled debate over the origin of life. Not visible to the naked eye, these microfossils were usually examined in thin sections under the electron microscope.

Paleontologists, like other biological scientists, came to depend more and more on the new observational techniques provided by improved microscopy. They also employed other laboratory techniques to document evolutionary change, to understand its progression, and to correlate the appearance and disappearance of certain fossil groups with changed environmental conditions on earth. It is precisely this combination of laboratory research with the traditional natural history tasks of naming, classifying, and describing new specimens that distinguishes the twentieth century. Accordingly, the chief museums became more than just local repositories for natural history collections; they grew into major research centers for investigation of the process of evolution.

The new techniques were, of course, applied not just to microfossils but to the fossils of larger organisms as well. As the century opened, Russian scientists excavated an unusually complete carcass of a mammoth in Siberia. It and other specimens were studied using methods from anatomy, histology, and biochemistry, as well as the more traditional descriptive techniques then characteristic of paleontology and geology. Remains of food found in the mouth and stomach, for example, were analyzed. The resulting data, along with studies of paleolithic paintings and sculptures of the mammoth, allowed biologists to reconstruct the structure and habits of these extinct animals. Expeditions on other continents also uncovered huge skeletons, including some of the mastodon, a larger elephantlike creature related to the mammoth.

Dinosaur fossils had also been studied in the nineteenth century, and finds of bones and tracks continued to attract attention and arouse debate in the twentieth. As with the mammoth and mastodon, careful reconstructions of these huge creatures graced the majestic halls of major natural history museums. Further discoveries, especially after mid-century, led to detailed examination of the dinosaurs' habits and to reassessment of their presumed relationship to other animals living during the same geologic era. Recurring and heated controversy surrounded the evolutionary relationships of dinosaurs, their physiological characteristics (including their possible warm-bloodedness), and their sudden mass disappearance from the geologic record. These debates raised some of the same theoretical issues that had been of central concern in the previous century.

Consider, for example, the sudden appearance or disappearance from the geologic record of entire fossil groups. Before the nineteenth century most geologists believed that only catastrophic changes in conditions over the earth's crust could account for modern landforms. Natural catastrophes (as well as divine acts) could also explain the extinction of species. But catastrophism lost favor. From the nineteenth through the mid-twentieth centuries uniform change over time was a central tenet of geological and evolutionary theory. Then, in the latter part of the century, fresh data appeared, and catastrophism reemerged as a subject of debate. Many species died out at the end of the Mesozoic era, about 65 million years ago. Abnormally large traces of iridium, an element rare on earth but more abundant in extraterrestrial bodies, were found in sediments of about that age. Some scientists argued that the collision of a comet or asteroid with the earth might provoke the kind of widespread environmental changes that could cause the extinction of entire populations and/or species.

The debates over evolution brought to bear not only the paleontological record but also data from embryology, genetics, and ecology. The development of each of these biological disciplines around parallel but overlapping sets of problems produced in the second quarter of the century a "synthetic" theory of evolution that drew on experimental and biochemical evidence, as well as on the anatomical record as reconstructed by paleontologists. Paleontology was not alone: during the century the changing emphasis of these other major disciplines that were derived from the natural history tradition (that is, embryology, genetics, and ecology) also reflected the crucial role of evolutionary theory in shaping the questions researchers put to nature. Chapter 11 will explore the major implications for evolutionary theory of the synthesis of these several related lines of research.

93. Tools used by the paleontologist.
Here, in a photograph from the mid-1930s Scarritt Expedition of the American Museum of Natural History, we see such field tools as picks, hammers, brushes, and supplies for labeling and record keeping. By carefully uncovering and labeling fossil-bearing rocks, researchers can document the appearance of specific groups of organisms in the strata.

94. Petrified forest. Lying on the desert floor in central Arizona are fossilized trees, whose natural wood fibers have been replaced by minerals. Scientists reconstruct the characteristic landscape of past geological eras by studying groups of fossilized plants and animals found preserved together. Analysis of evolutionary relationships has stimulated much of this work in the twentieth century.

95. Excavating the Berezovka mammoth. In 1901-1902 an expedition of the Imperial Saint Petersburg Academy of Sciences excavated an extraordinarily complete set of 45,000-year-old mammoth remains on the bank of the Berezovka River in northern Siberia. This view, from an early stage in the excavation, reveals the mammoth's right foreleg and skull.

96. Berezovka mammoth on exhibit. A restored version of the mammoth was mounted, exactly as the animal lay when discovered. It is now housed in the Zoological Museum of the Soviet Academy of Sciences in Leningrad. Microscopic and biochemical analysis of stomach and mouth contents enabled scientists to reconstruct the dietary habits of these extinct relatives of the elephant.

97. Exploration in the Gobi. Members of the American Museum of Natural History's Central Asiatic Expedition of 1930 here excavate three jaws of the shovel-jawed mastodon, *Platyklodon*. The museum's expeditions to Central Asia after World War I were extraordinarily productive. Major discoveries included the first known dinosaur eggs and fossils of previously unknown reptiles and mammals, as well as signs of prehistoric human life. The expeditions were led by the colorful naturalist and explorer Roy Chapman Andrews, later director of the museum. Andrews, whose adventures included gunfights with Mongolian bandits, reportedly was the model for the movie hero Indiana Jones.

98. Simpson excavating an Eocene skeleton. On a
1930-1931 expedition to Patagonia, George Gaylord
Simpson, perhaps the preeminent American paleontologist of his time, found the most complete mammal
skeleton from the Eocene epoch (54 million to 16 million years ago) ever discovered in South America. The
moderate, middle-sized mammal had been named
Thomashuxleya externa in 1901 on the basis of only a jaw
fragment.

99. At work on an *Eohippus* block. This 1959 scene
suggests the careful, painstaking work required to free
a specimen (in this case, an Eocene ancestor of the
horse) from the surrounding rock. Note the hammer
at center and the several brushes used to remove the
particles of dust and rock from the specimen.

100. Dinosaur tracks. The dinosaur trail along the Paluxy River near Glen Rose, Texas, was discovered in 1938 and subsequently excavated. A four-footed sauropod dinosaur produced the large tracks; the hind feet of a bipedal carnivorous dinosaur the smaller three-toe ones. The tracks were made in soft sand in the late Jurassic period, about 135 million years ago.

101. Workroom of a paleontologist.
The room belonged to Henry Fairfield Osborn, who served as president of the American Museum of Natural History from 1908 to 1933. He was largely responsible for building the institution into the biggest natural history museum in the world. Among the fossil specimens visible is the large mammal *Palaeosyops*, a tapir-like titanothere.

102. Preparator's room. Exhibit specimens are reassembled and prepared for display in this workroom at the American Museum of Natural History.

103. Preparing a fossil. Here we see Otto Falkenbach at work in the American Museum in 1925 reconstructing the shell of a fossil giant turtle. The oldest rocks in which turtles appear date from the Triassic period (225 million to 190 million years ago) in the early Mesozoic era.

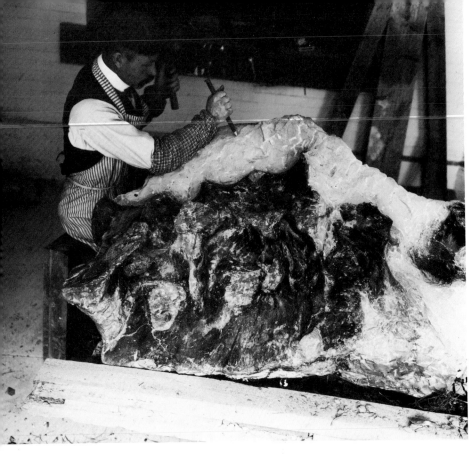

104. Dinosaur bones in block. The block contains the pelvis and sacrum (lower part of backbone) of the huge carnivorous dinosaur *Tyrannosaurus rex*. The bones are from the Mesozoic era, also known as the Age of Reptiles. As much as 47 feet long and 19 feet high, *Tyrannosaurus rex* walked on only two legs. The first *Tyrannosaurus rex* skeleton was discovered in Montana in 1902 by the noted American fossil hunter Barnum Brown.

105. Assembling a skeleton. The man at work in this 1936 scene is assembling bones of a dinosaur, probably *Nodosaurus*. The discovery and mounting of dinosaur skeletons was a very active branch of paleontology early in the century. As more and more species were recognized after mid-century, dinosaurs' physiology and evolution became a major subject of debate.

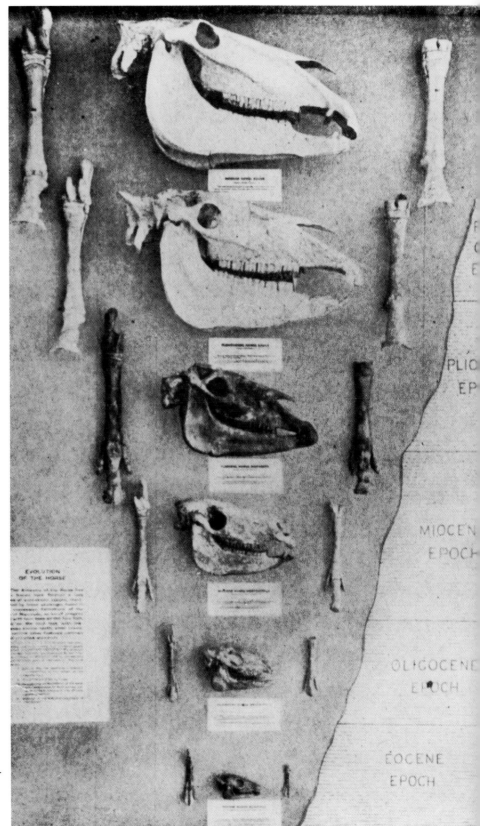

106. Evolution of the horse. This museum exhibit traces the line of descent from *Eohippus* in the Eocene epoch to *Equus* in the modern period. To the left of each skull is a hind foot; to the right, a forefoot. Scientists identified the line of descent of the contemporary horse from its tiny ancestor by studying changes in dentition, length of head and neck, leg structure, and the fusion of toes to form a hoof. The century saw paleontologists become increasingly interested in understanding the direction and rate of evolution. Such reconstructed lines of descent were used in debates over how evolution occurs.

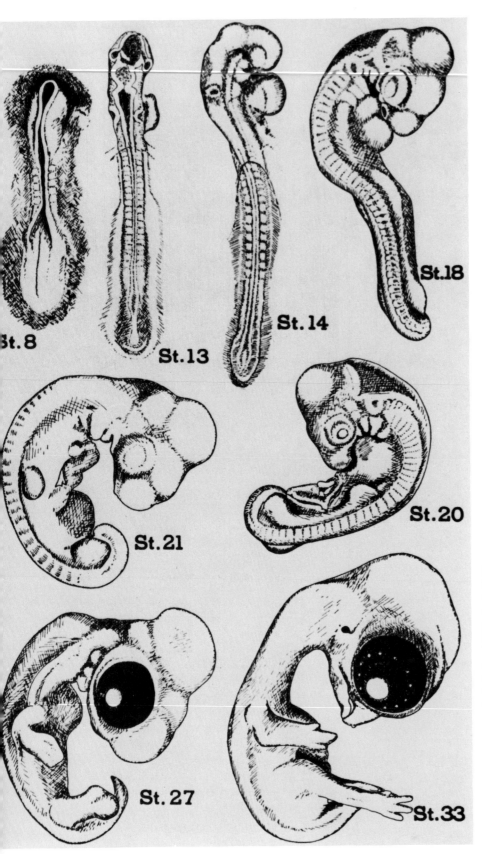

St. 8

St. 13

St. 14

St. 18

St. 21

St. 20

St. 27

St. 33

107. Stages in the development of the chick. Traditionally concerned with describing developmental stages, embryologists in the late nineteenth and early twentieth centuries turned in large measure to experimentation. They manipulated growing animal embryos by altering the growth conditions and studied the results. They did not, however, abandon simple observation of normal development. These drawings of the chick are mainly from mid-century. While the development of the chick egg has been intensively studied since ancient times, the twentieth century saw simple marine animals, easily accessible at the newly established biological stations, increasingly used for research on the factors influencing development.

CHAPTER

8

*E*mbryology

The continuity between organisms past and present is maintained through the processes of inheritance and development, in previous centuries grouped together under the term "generation." Embryology is the part of this subject dealing with the development of immature forms of organisms into the adult. It has been a major focus of biological research since ancient times.

Embryologists have traditionally observed and described change in form over time. But in the twentieth century they resorted to an additional approach to studying development: experimental manipulation and analysis. Besides observing and carefully describing each stage of embryological development, they began to investigate precisely how development occurred. They started to explore the unknown territory between heredity and physiology—that is, how it is that offspring come to resemble their parents as they grow. This research focused increasingly on the cell. Watching the egg and sperm combine at fertilization and transform into complex, organized, multicellular forms, embryologists began to study the individual parts of the cell, like the nucleus and chromosomes, and their particular roles in these processes.

The chick was the major traditional subject for developmental research. But beginning in the late nineteenth century it increasingly had to share pride of place with a variety of marine organisms, whose shell-less embryological transformations could be more easily observed. Much of this work was conducted at the seaside in the marine laboratories that proliferated at the end of the century. These studies reflected newly refined goals tied to both evolutionary and physiological theory.

First of all, the theory of evolution strongly encouraged study of the similarities between the embryonic forms of different species. Embryologists took advantage of the improved optical microscope to observe the developmental forms of a diverse array of invertebrate organisms, including sea urchins, sponges, starfish, and marine worms. They continued this descriptive work with a fresh vigor in the twentieth century.

Second, embryologists became experimentalists, building on nineteenth-century observations of chromosomes' behavior during cell division, the process of fertilization, and the fate of specific cells during cleavage (the transformation of the single-celled egg into a multicelled embryo) and embryonic development. They also explored the regenerative capabilities of selected organisms. In their experiments they subjected embryos to altered chemical environments and began to probe the biochemical basis of the embryo cells' behavior. They observed the effect of cutting or tying off one group of cells in the embryo from others. Such micromanipulative studies permitted analysis of the mechanics of fertilization, development, and cell determination (or fate) in a variety of organisms.

Experimental manipulation of the developing

embryo allowed researchers like the German embryologist Hans Spemann to assess how the form and function of specific embryonic cells are determined. Eventually, experimental manipulation permitted them to evaluate the extent to which factors in the nucleus, cytoplasm, or cell environment control development. As the century progressed, it became clearer and clearer that many developmental events are chemically controlled.

Using not only newly available biochemical techniques but also moving pictures and the electron microscope to observe development ever more closely, embryologists contributed extensively during the century to the growth of the more general area of reproductive biology. This area attracted special interest in the middle decades, in part because reproduction and development also had become important areas of investigation for physiologists. Hormones, potent chemicals that affect growth and development, were found to be produced by a variety of tissues in the body. Combining techniques from microscopy, embryology, biochemistry, and physiology made it possible to analyze oogenesis and spermatogenesis (formation of the egg and sperm), ovulation, fertilization, cleavage, and embryonic development.

Emphasizing the relatedness of all creatures and uncovering common mechanisms and structural features that underlie development, embryologists came to focus on the basic problem of function as well as form. Meanwhile, continued application of new chemical and physiological techniques opened the door to yet another set of questions—the molecular basis of development. This area became fruitful in mid-century through study of the reproduction and development of very simple organisms—molds, bacteria, and viruses. Biochemical analysis of the developmental processes of these microorganisms prompted unified research programs in the latter decades, introducing molecular explanations for evolutionary processes while at the same time explaining basic biochemical processes in the cell and in the organism.

Embryology in the twentieth century thus has dealt with the interface between macroscopic and microscopic biological events and furnished the common ground unifying the separate natural history and physiological traditions of previous centuries. The shift from the study of form to the study of functional phenomena can be seen clearly in embryologists' changing research interests. As the century progressed, their research domain expanded from morphology and altered form, the traditional purview of embryology, to the conceptually larger domain called developmental biology. One textbook series published in the 1960s explained this enlarged focus, which reflected a trend in all the life sciences: "Major functional phenomena rather than catalogues of animals and plants comprise the core of Modern Biology. Such heretofore unrelated fields of cytology, biochemistry, and genetics are now being unified into a common framework at the molecular level." By its very title one of the texts in the series, *Interacting Systems in Development* (published in 1965), underscored the unification of the embryological, physiological, and genetic perspectives that had occurred in embryological studies—part of the methodological and conceptual unification of the life sciences that accompanied the rise of the experimental attitude in biology.

108. Experimental study of development: Delage's laboratory. The French biologist Yves Delage headed the Roscoff Biological Station early in the century. His research on marine invertebrates included contributions to the study of fertilization, artificial parthenogenesis, and the embryology of sponges.

109. Observing embryological development: Just. One of the first American blacks to achieve prominence in zoology in the early decades of the century, Ernest Everett Just was responsible for fruitful research on the physiology of development, fertilization, and cell division. Just, a professor at Howard University in Washington, D.C., worked both at the Marine Biological Laboratory in Woods Hole, Massachusetts, and the Zoological Station in Naples, Italy.

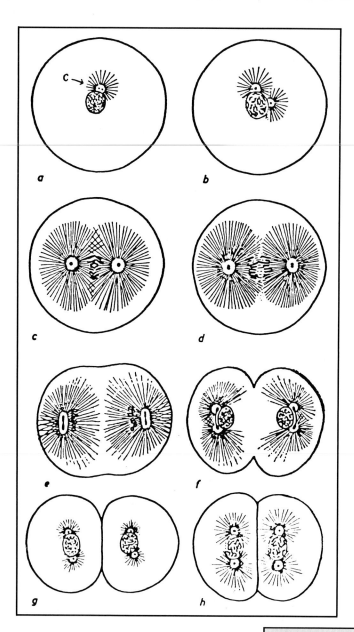

110. First cleavage division of the sea urchin egg. Sea urchin development was a favorite subject for study in the late nineteenth and early twentieth centuries. These semi-schematic drawings reflect the observations of the German biologist Theodor Boveri. In addition to making key contributions to understanding of the role of the nucleus in development, he showed that the tiny structure called the centrosome is the division center for the dividing egg cell. The drawings show the cell's nuclear material being divided and the subsequent formation of two daughter cells, which also prepare to divide. In the initial drawing the centrosome, surrounded by astral radiations, is indicated by *c*.

111. Parthenogenesis. A turn-of-the-century experiment by the German-born American biologist Jacques Loeb seemed to give weight to a mechanistic interpretation of life. The cell division of sea urchin eggs ordinarily triggered by fertilization was initiated without sperm. An egg in a solution of saponin and seawater (*top left*) formed a membrane (*top middle*) indicating that the egg was fertilized. If the egg was then washed and put in strong seawater, it developed normally. If this was not done, the protoplasm dissolved (*bottom right*). · The role of the sperm cell in fertilization would be elucidated decades later by studies in genetics and cell biology.

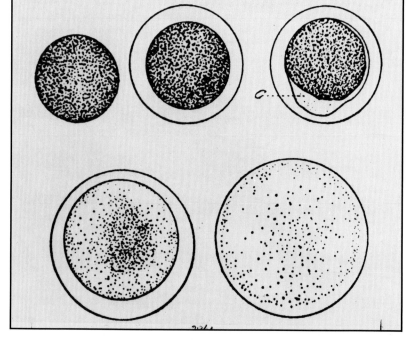

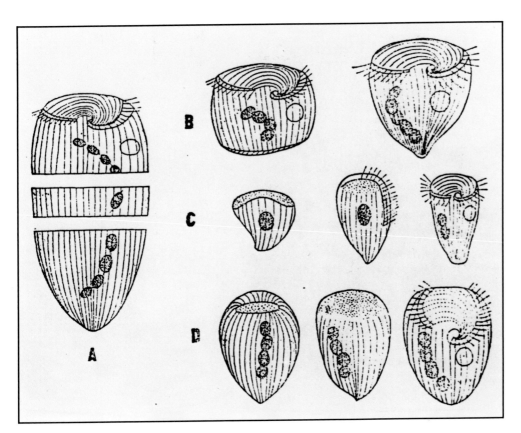

112. Regeneration. The type of protozoan (one-celled animal) known as *Stentor,* shown here in a diagram originally published around the turn of the century, was one of the favorite subjects of experimentation for researchers studying the ability of a fragment to regenerate an entire organism. In **A** the protozoan is cut into three pieces. **B**, **C**, and **D** show the results of the experiment. Each of the three pieces is capable of regenerating the organism.

113. Experimenting with chick embryo development. By interfering with the development of the embryo scientists could track the fate of specific cells. These drawings depict the fate of the "embryonic disk," or blastoderm, when different parts of the blastoderm were injured with a hot needle. The injuries could be recognized in later developmental stages and provided a means of tracing what develops from what. In the drawings the horizontal line cuts through the area where the head joins the neck. The head develops from the front (upper) half of the disk.

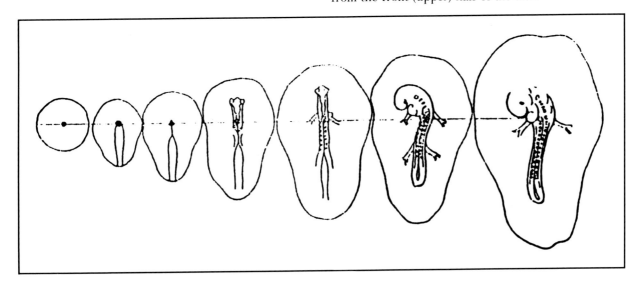

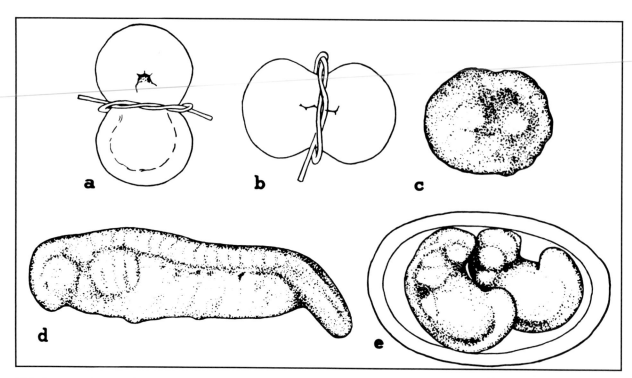

a   b   c

d   e

114. Spemann's experiments. The
German embryologist Hans Spemann
found that when he tied a freshly
laid newt egg and constricted it into
an hourglass shape, the egg sometimes developed into two complete
larvae, and sometimes into a larva
and a mass of belly tissue. Spemann
suggested that a dark area in the egg
cytoplasm called the gray crescent
controlled development. When the
constriction cut through the gray
crescent, both halves of the egg developed normally. If the gray crescent was located entirely in one egg
half, only that half developed normally; the other formed a "belly." In
this drawing from Spemann's 1938
paper, the constriction in the frontal
plane (**a**) leads to formation of a
"belly" (**c**) from the ventral half and
a complete embryo (**d**) from the dorsal half. In contrast, medial constriction (**b**) leads to the formation of two
complete embryos (**e**). Spemann
showed that both nuclear and cytoplasmic elements were required for
normal development.

115. Transplantation, induction. In the mid-1920s
Spemann and his student Hilde Mangold found that
by transplanting tissue from the "dorsal lip" of one
amphibian embryo to the flank of another, they could
induce development of a neural tube, an embryonic
structure capable of forming the brain and spinal
cord. Spemann suggested that an "organizer"
substance might be responsible.

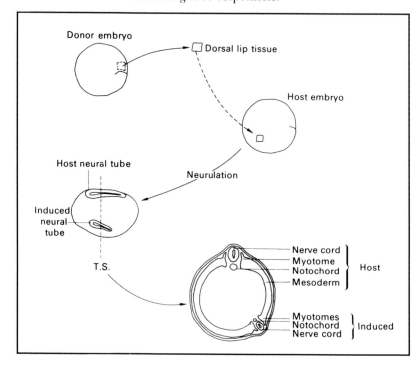

116. Limb induction by transplants. The Russian-born zoologist Boris Balinsky succeeded in inducing the growth of an extra limb on a salamander by grafting an organ rudiment onto the embryo. Here we see the effect of grafting a nose rudiment from embryo **a** onto embryo **b**. The extra limb is shown in **c**. Experiments such as these suggested how the orderly process of development might be regulated in the normal embryo.

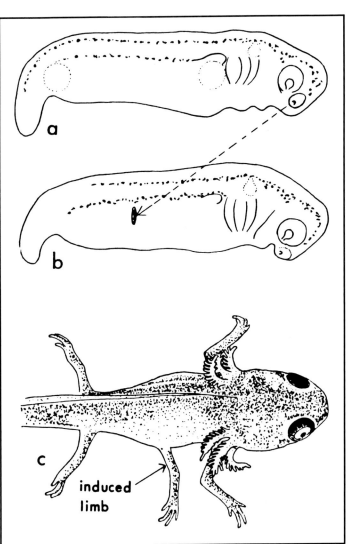

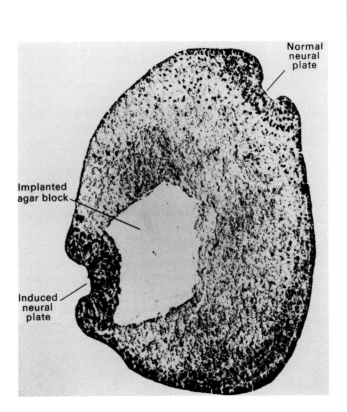

117. In search of the organizer. Spemann's collaborator Wehmeier implanted an agar block containing tissue extract into a salamander embryo. The implant induced a neural plate (or incipient neural tube) above it. Such experiments suggested that a diffusible chemical regulator might be responsible for the effect.

118. Ovulation. The release of the egg from the ovary was an important area of research in the 1930s. These time-lapse pictures published in 1935 show the process in a rabbit. First we see two follicles one and a half hours (**1**) and thirty minutes (**2**) before rupture. A clear fluid is exuded in the early phases of rupture (**3**). Later the fluid shows some blood in its mass (**4,B**). Another follicle becomes conical as rupture approaches (**4,A**). This new follicle also begins to rupture (**5**). Eight seconds later there is a final gush of fluid which carries the ovum away from the ovary (**6**). The final frames show the ovum surrounded by follicle cells in the follicular exudate (**7**) and isolated in a special preparation (**8**).

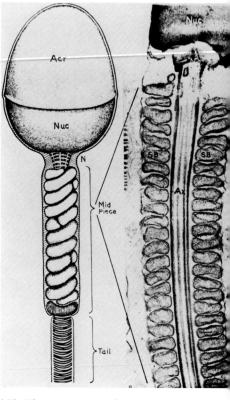

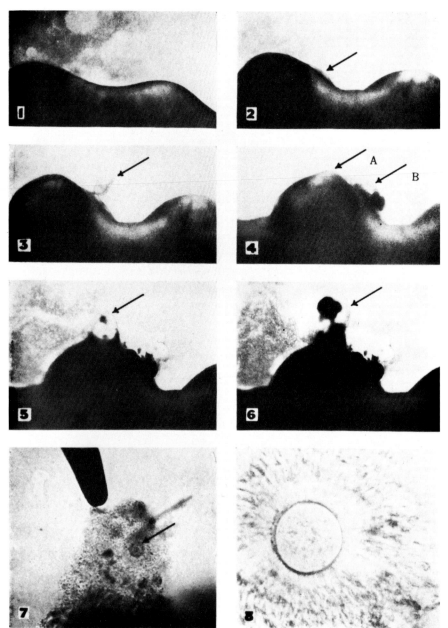

119. Fine structure of spermatozoa. Knowledge of sperm physiology, so critical to the understanding of fertilization, was enhanced by the ultrastructural studies made possible by the electron microscope. Above, at left, is a mid-century diagram of a human sperm; at right, an electron micrograph of a bat spermatozoan. Elements that can be seen here include the acrosome (Acr), which releases egg-penetrating enzymes; nucleus (Nuc), containing the hereditary material; neck (N); axial filament (AX), surrounded by the axial sheath; spiral body (SB); and tail piece (T).

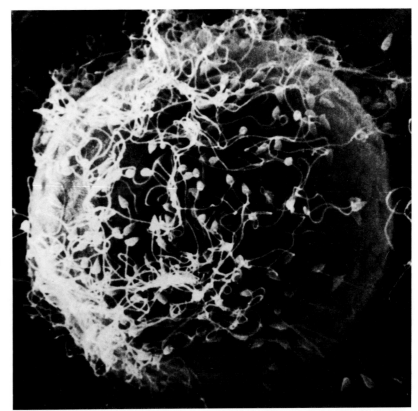

120. Sperm and egg prior to fertilization. How is the egg stimulated to develop into a multicellular organism? Numerous theories have been put forward. This scanning electron micrograph published in the 1970s shows sea urchin sperm binding to the surface of an egg. Only one sperm will succeed in fertilizing the egg. The plasma membrane, the outer membrane of the egg cell proper, is below the surface envelope, or vitelline layer, seen here.

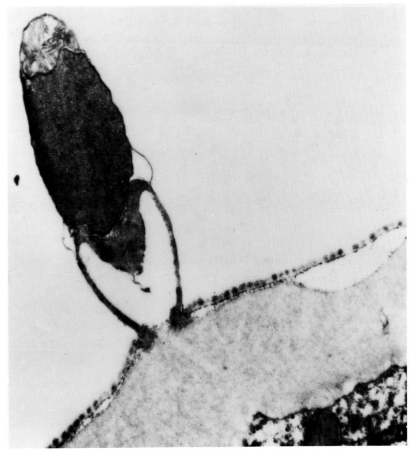

121. Sperm and egg: initial contact. The plasma membrane of the sperm meets the envelope of the egg. Subsequently, the nuclear material of both the sperm and egg combine, and the fertilized egg develops. The process of fertilization and its developmental consequences are central areas of twentieth-century research. Developmental events have been studied by cytological, embryological, genetic, and biochemical techniques.

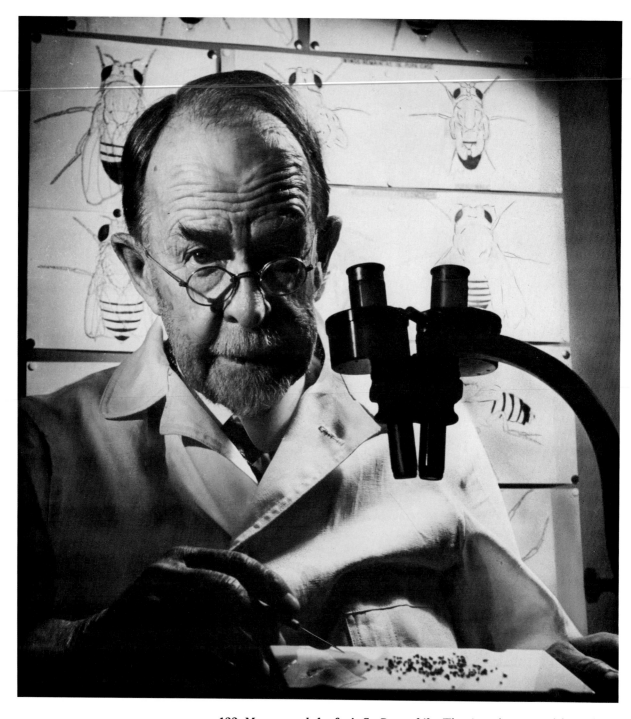

122. Morgan and the fruit fly *Drosophila*. The American geneticist and embryologist Thomas Hunt Morgan won the Nobel Prize in physiology or medicine in 1933 for his work on heredity in the fruit fly *Drosophila*. Featuring a short life span and rapid succession of generations, *Drosophila* provided a relatively simple experimental system with which Morgan and his colleagues could establish the chromosome theory of heredity. Genes, Morgan's group showed, are arranged in a linear order on chromosomes in the nucleus, and individual genes are associated with identifiable hereditary traits.

9

*G*enetics

Genetics is a young science. For generations observers had noted that traits are inherited, but the phenomenon was not thoroughly documented and analyzed until the twentieth century. Correlation of the inheritance of traits with the behavior of visible elements (chromosomes) within the cell nucleus led scientists to identify and map new biological units called genes, confirming the material basis of heredity. Later in the century genes were found to be a kind of code in the molecular structure of nucleic acid. This discovery revolutionized biological research, uniting biochemical, physiological, and evolutionary biology into one coherent theoretical framework based on the analysis of molecular processes.

Biologists have always been preoccupied with the continuity of form in living things. But before the twentieth century agricultural practices as well as biological explanations of structure rested largely on the simple observation that like begets like and that children resemble their parents. Systematic, experimental study of inheritance did not begin until the nineteenth-century debate over evolution. Even then the value of such research was not immediately recognized. The monk Gregor Mendel's experiments with the garden pea in the late 1850s to 1860s are now famous, but they were overlooked at the time. Mendel's work was rediscovered in 1900 by several scientists studying the problem of inheritance. A major question at the turn of the century was how the variation necessary to any progressive evolutionary process arose.

Research with crossbreeding, such as Mendel's, suggested that factors from each parent were combined in the production of each new generation. Both plant and animal studies supported Mendel's interpretations.

Plant and animal genetics developed rapidly in the first two decades of the century. Research on inherited traits in human populations, however, provoked much controversy, especially before World War II. The distinction between hereditary and environmentally determined characteristics (nature versus nurture) was unclear, and geneticists debated how best to collect and evaluate human data. Attempts by some governments to apply the principles of genetics to human populations in order to direct the evolution of human stock—the practice of eugenics—led to polemics over the political uses and abuses of limited scientific knowledge. The direction of human evolution became a key item on agendas for social reform during those years, and the Nazis justified genocide with "biological" doctrines.

In the midst of these debates laboratory study of *Drosophila melanogaster* provided a relatively simple way of examining how traits are inherited in animals and how variation arises. This tiny fruit fly had the advantage of breeding rapidly. The American biologist Thomas Hunt Morgan and his colleagues used it to investigate the inheritance of external characteristics like eye color and wing form, noting the sudden appearance of new characteristics or mutations and correlating these observations with the behavior of chromosomes in the animal's cells. Morgan's group firmly established the material basis of heredity. They devised experiments to prove the notion that genes are real material entities. They produced maps of the location of specific genes on individual chromosomes. Looking at the banding produced by staining the large chromosomes found in *Drosophila* salivary glands, they saw that chromosomes displayed a "definite and constant morphology." Cytological study of the chromosomes of other organisms showed a similar regularity of structure. The mapping of chromosomes and the analysis of mutations thus became an important focus for cytogeneticists, who investigated inheritance using histological techniques. Abnormal characteristics in the adult animal could be read from visible alterations in chromosome structure.

Work on the genetics of agriculturally important plants like corn and wheat ultimately led to the production of new commercial varieties. In the first half of the century field research on crossbred plants was analyzed both by traditional breeding methods—that is, by the yield from any given variety—and by newer cytogenetic techniques. Careful studies of these phenomena by the American geneticist Barbara McClintock led finally to the discovery of the transposable gene elements called jumping genes.

In the 1920s geneticists began to see that radiation could induce sudden changes, or mutations, in genes. Herman J. Muller of Morgan's group was among the scientists who first studied this phenomenon. After World War II, which saw the development and use of the atomic bomb, researchers devoted more attention to the effects of radiation on human populations. They recognized that some of the genetic effects could be harmful. Radiation from a variety of sources was later used to study the origin and inheritance of mutations, particularly among crop plants.

In the 1930s and 1940s the study of inheritance in microorganisms began to provide important evidence linking genetic and biochemical phenomena. This research, together with chemical analyses of nuclei and chromosomes, suggested a chemical basis for observed genetic phenomena. The story of this emerging synthesis, which burst into prominence with the rise of molecular biology in the 1950s, is told in Chapter 16. Drawing on data from X-ray crystallography, biochemical genetics, and studies of the biology of the class of viruses called bacteriophages, molecular biology led to a whole new understanding of the unit called the gene.

123. Mendel's garden. In this monastery garden measuring only 15 feet by 30-40 feet, located in Brünn, Austria (now Brno, Czechoslovakia), the nineteenth-century monk Gregor Mendel carried out extensive experiments in hybridization, crossing tens of thousands of pea plants and studying the inheritance of specific traits, including flower and seed color, stem length, and seed coat texture. Analysis of the crosses led him to discover the law of segregation and the law of independent assortment, the principles of heredity now known as Mendel's laws.

124. Mendel's manuscript. This is the first page of "Experiments With Plant Hybrids," a paper published in 1866 in which Mendel reported his research and conclusions. The paper was virtually ignored until 1900, when it was discovered by three scientists also working in plant hybridization: the Dutchman Hugo de Vries, the German Carl Correns, and the Austrian Erich Tschermak von Seysenegg. The rediscovery of Mendel's work led to intense controversy over the nature of variation in organisms. His discussions of the properties of hereditary factors (later named genes) correlated well with the observed behavior of chromosomes during cell division.

125. De Vries and the evening primrose. Drawing largely on experiments with the evening primrose, *Oenothera lamarkiana*, Hugo de Vries attributed evolutionary change to sudden changes, or mutations, that produced new species in a single generation. His mutation theory was popular early in the century because it seemed to explain problems left unanswered by Darwinian theory. But the "mutations" de Vries observed in *Oenothera* were actually due to recombinations of already existing characters. Nonetheless, he helped establish the tradition of experimental analysis in studies of evolution.

126. Spillman's crosses of wheat. The American agricultural economist William Jasper Spillman carried out research on plant heredity in the late nineteenth and early twentieth centuries. Agriculturalists at experiment stations were not the only ones to develop new varieties of plants through pedigree selection, hybridization, and backcrossing. Private breeders also did this. In fact the early twentieth century saw breeders provide substantial support and data for the newly emerging science of genetics.

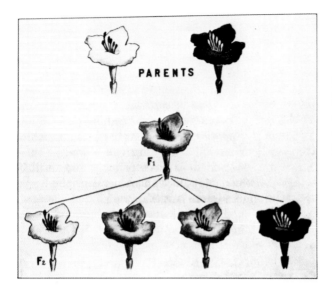

127. Crossing a red and white four-o'clock. This diagram is from the enormously influential book *The Mechanism of Mendelian Heredity*, published in 1915 by T. H. Morgan and his students A. H. Sturtevant and C. B. Bridges. The authors drew extensively on the data from their *Drosophila* research correlating the results of breeding experiments with observations of the behavior of chromosomes in the cell. In the diagram a red four-o'clock (*Mirabilis jalapa*) is crossed with a white one, producing a pink offspring, which possesses genes for both red and white flowers. These show up in the next cross. Morgan and his associates produced evidence that the dominant and recessive genes revealed by such crosses were actually physical units located on the chromosome.

128. Zebra hybrid with mother. Animal breeding was another important source of data for theories of heredity in the late nineteenth and early twentieth centuries. Here are seven-day-old Romulus and his dam, Mulatto, a black Highland pony. The Scottish zoologist James Cossar Ewart used them in crossbreeding experiments. Romulus's father was the zebra stallion Matopo. Ewart disproved the ancient theory of telegony, which held that the heredity of an individual is influenced not only by the father but also by the previous mates of the mother. The photograph is from Ewart's book *The Penycuik Experiments*, published in 1899.

129. Eugenics Record Office. The British anthropologist Francis Galton argued that the human species could control its evolutionary future, improving itself through selective parenthood. In 1908 he founded the Eugenics Education Society. Similar organizations were established in other countries. A key institution in the American eugenics movement was the Eugenics Record Office at Cold Spring Harbor, New York, which collected and stored family histories and "human pedigrees."

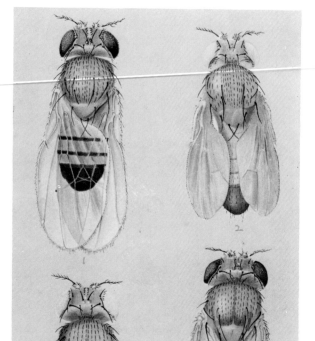

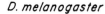

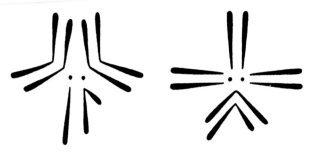

130. *Drosophila* **chromosomes: two species.** Fruit fly species differ in size, body and eye color, food preferences, breeding site, and geographical distribution. The X and Y chromosomes form the bottom pair in each of these two diagrammatic representations.

131. Eye and wing mutations. Here is a plate from one of Morgan's early papers on *Drosophila melanogaster* (1911). Morgan found that certain traits were associated with particular chromosomes. The inheritance of white eyes, instead of the normal (wild-type) red eyes, for example, was linked to the X chromosome. Of the four flies shown, the first is a normal wild-type fly with red eyes and a dark body. The next two are white-eyed with miniature wings, the second having a yellow body and the third a dark body. The fourth is red-eyed but has a brown body and longer than normal wings.

132. Bateson and Morgan's group. In 1921 the English Mendelian William Bateson, a leading opponent of Morgan's chromosome theory, visited the Morgan group at Columbia University. Discussions in the "fly room," especially with C. B. Bridges (*far right*) and A. H. Sturtevant, finally convinced Bateson (*seated*) that chromosomes actually bear the Mendelian genes.

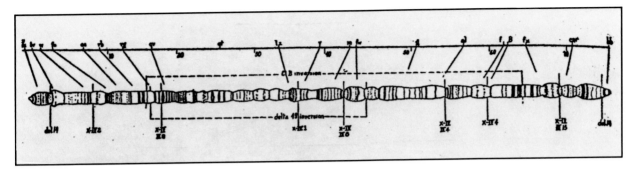

133. Pioneering chromosome map.

The American geneticist T. S. Painter of the University of Texas published this drawing in 1933. He made it with a camera lucida, which permitted him to trace sections of the microscopic image of the chromosome of *Drosophila melanogaster*. Most chromosomes are extremely tiny, but Painter overcame this difficulty by using the unusually large chromosomes found in the *Drosophila* salivary glands. He stained the giant chromosomes with acetocarmine to obtain the banded pattern enabling precise determination of the gene locations shown here. This was the first time that the postulated linear arrangement of genes on chromosomes had actually been physically mapped. Previous mapping attempts had merely calculated the relative positions of various genes from breeding data.

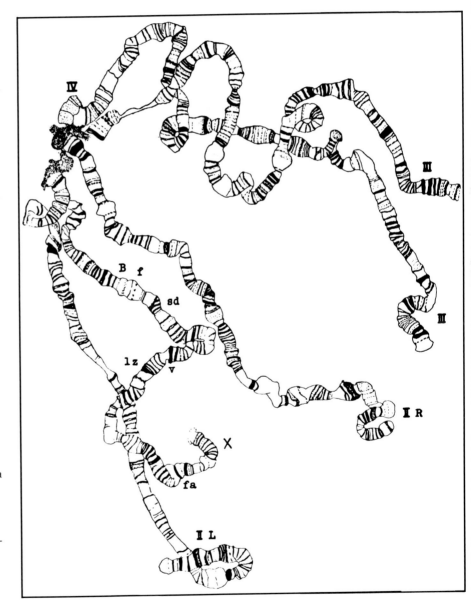

134. *Drosophila* giant chromosome.

In a camera lucida drawing that was published by Painter in 1934, we see the complement of chromosomes in a cell from the salivary gland. These chromosomes are a hundred times larger than those found in the germ (reproductive) cells. By using such giants, researchers could see and precisely map structural alterations—like inversions, deletions, and translocations—in the position of genes on the chromosomes.

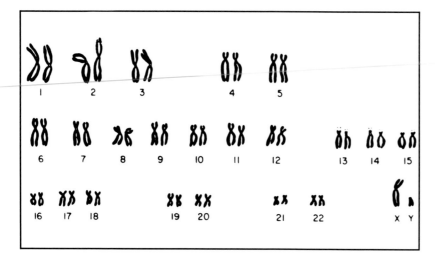

135. Human karyotype. Here, tidily grouped and numbered according to an established convention, are the chromosome pairs of a normal male. The male differs from the female in having an X and a Y chromosome instead of two X chromosomes. A male with Down's syndrome would have three rather than two chromosomes numbered 21.

136. Dividing cells in a growing onion root tip. Chromosomal movements were observed by nineteenth-century scientists with the aid of the microscope. These gained new meaning when Mendelian "factors," or genes, were identified with the chromosomes. The cells in this turn-of-the-century drawing show chromosomal formations typical of mitosis, or cell division. Nondividing cells (a) display the chromatin network in the nucleus. Some nuclei prepare for division (b) while others are observed to be dividing (c). Recently produced daughter cells may also be seen (e).

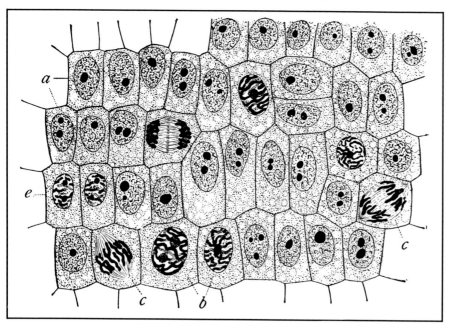

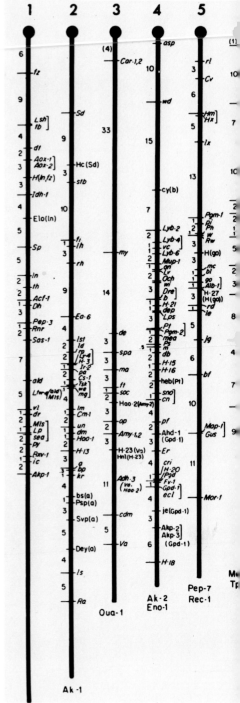

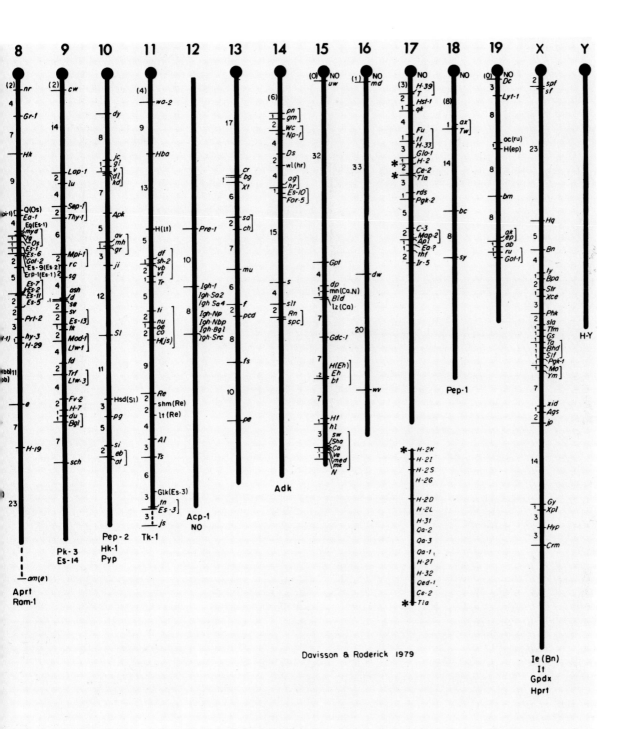

137. Mouse chromosome map. This "linkage" map shows the relative positions of the known genes on the mouse chromosomes. Genes located on the same chromosome tend to be inherited together. The map was prepared at Jackson Laboratory (Bar Harbor, Maine) in 1979, the year that institution celebrated its fiftieth anniversary. The map represents 64 years of research; 423 gene symbols are indicated. There is a detailed analysis of one section of chromosome 17, indicated by asterisks.

138. Corn rows at Cold Spring Harbor. The American botanist and geneticist George Harrison Shull worked at the Carnegie Institution's Station for Experimental Evolution at Cold Spring Harbor, New York, from 1904 to 1915. His experimental studies of corn led to the development of the concept of hybrid vigor, or heterosis, the increased yield and growth that can come from cross-breeding unrelated strains. Shull was one of the first to isolate pure strains of corn by inbreeding and to cross them to produce corn showing hybrid vigor. His work and that of Donald F. Jones of the Connecticut Agricultural Station and Edward M. East of Harvard's Bussey Institution laid the foundations for the development of hybrid seed.

139. From the golden age of maize cytogenetics. In the late 1920s and early 1930s cytogenetic work with maize accomplished for plant genetics what had already been achieved for animals by *Drosophila* studies. This 1929 photograph, taken at a lunchtime seminar in the corn hollow at Cornell University, includes George Beadle (*kneeling*) and Barbara McClintock (*far right*), both of whom would later earn Nobel Prizes. McClintock's work on maize led her to the discovery of "jumping genes," mobile or transposable genetic elements. Beadle's work with the microbiologist Edward Tatum led to the idea that genes produce their effects by coding for enzymes, which in turn catalyze the biochemical reactions that produce observable traits. Others in the picture are (*left to right*) Charles Burnham, Marcus Rhoades, and Rollins Emerson.

140. Atomic farm. The effects of radiation on plants are studied at this "Gamma Garden" at Brookhaven National Laboratory in New York. Radioactive cobalt, raised up into the central vertical pipe from an underground storage pit, irradiates plants growing at various distances. The gamma rays given off by the cobalt can produce mutations, which may prove useful in breeding improved agricultural crops. The mutagenic effects of radiation have been studied since the 1920s, when Hermann J. Muller, a member of the original *Drosophila* group under Morgan at Columbia, investigated the effect of X rays on *Drosophila*—research that won him the Nobel Prize in physiology or medicine in 1947.

141. Radiation-induced mutation. The male parents of these *Drosophila* were given food containing radioactive phosphorous. The radiation produced a genetic change in the reproductive cells causing subsequent generations to have the rudimentary wings seen here.

Mr Blackman and Mr Tansley.
General Botany (Intermediate).
Times to be arranged. £1. 1s.

142. Blackman and Tansley. This caricature was made by D. G. Lillie in 1908, when both Frederick Frost Blackman (*left*) and Arthur G. Tansley were at Cambridge University. Tansley founded the journal *New Phytologist* in 1902 and helped establish the *Journal of Ecology* in 1913. He produced classic works on vegetation and, in 1935, introduced the term "ecosystem." Blackman, Tansley's brother-in-law, was best known for his work on plant physiology, particularly photosynthesis. Both men were enormously influential teachers.

10

Ecology

Ecology deals with the interrelations between organisms and their environment. It has been called both the economics and the sociology of plants and animals. The distribution of organisms within the environment is a fundamental concern of the natural history tradition, from which much of twentieth-century biology grew. The British biologist Charles Elton, who helped found the science of animal ecology, said that the distinguished nineteenth-century German naturalist Alexander von Humboldt may have been the first ecologist, since he "created a stirring picture of the plant and animal world as a whole, with its majestic settings and its complex interplay of forces." Plant geographers did pioneering work in the nineteenth century. In the twentieth the tradition of classifying and describing flora and fauna—the domain of taxonomy—was supplemented and extended as biologists paid increasing attention to how organisms are distributed over an area and are interrelated.

The development of ecology as a distinct science was intimately related to the rise of genetics and evolution as separate disciplines in the early part of the century. Spurred on by evolutionary questions, researchers began systematically examining the relationship between genetic variation and environmental variation within specific geographical regions. Field ecologists studied environments and the distribution and abundance of species within them. They analyzed the composition and stability of communities of organisms over time. The study

of zonation in plant communities led from analysis of specific aggregates of plants and animals to documentation of sequential stages, or succession, in the development of specific communities, such as those found in ponds, meadows, and forests. Among the pioneers in succession research was Victor Ernest Shelford, who in 1916 became the first president of the Ecological Society of America.

Ecologists also began to evaluate the competitive relationships existing within communities, especially the predator-prey relationship so central to Darwin's theory of natural selection. Descriptive study of animal and plant distribution was eventually overshadowed by experimental analysis of the effects of competition and predation within communities.

Competition among individuals and species for essential resources could be investigated with particular precision in the laboratory, whose controlled conditions facilitated mathematical analysis and modeling. In the 1930s the Soviet scientist Georgi Frantsevich Gauze studied the growth curves of pure and mixed populations of organisms such as paramecia within an artificial environment. The principle that two species cannot occupy the same ecological niche at the same time is generally attributed to him, along with the American zoologist Joseph Grinnell. This line of research was greatly extended in the 1940s by Thomas Park, a University of Chicago animal ecologist who worked with two species of flour beetles. He found that in a mixed population one of the two always became extinct; since plenty of food was provided, the two apparently competed for space.

Ecologists' adoption of the experimental approach, paralleling experimentation in the other emerging life sciences, provided fresh data with which to discuss change in communities over time. The emphasis on precision and quantification was characteristic of the century's developing "population biology," in which populations replaced individual organisms or types as the unit of study.

Ecologists' research into the interrelations among community inhabitants enabled them to identify food chains and determine the energy relationships among various levels of the chain. This kind of quantitative analysis, when complemented by chemical studies, led to understanding of the ways in which essential nutrients are recycled within the environment. It also raised the possibility of modifying those processes in given communities.

The study of interrelationships within communities and analysis of the habitat and niche increased appreciation of the dynamic balance existing in the natural world and laid the foundation for resource management. Ecologists argued that overgrazing of pastures and deforestation leached the soil and radically altered the communities of organisms inhabiting an area. The introduction or selective killing of a single organism might affect the entire community. Increased understanding of these effects led eventually to efforts to reclaim unproductive land and to repopulate damaged environments. For example, simple mineral deficiencies could be remedied. Barren regions could be transformed by inoculating the soil with specific organisms.

In later decades experimental field studies of the role of forests in maintaining the productivity of surrounding lands led to greater understanding of how to reclaim an unproductive area by introducing specific plant and animal communities into it. Such environmental management techniques, which had actually been initiated in the early part of the century, raised the public's expectations of science, as people presumed—without sufficient justification—that scientists could adequately judge how best to manage the environment for human ends.

143. Vegetation zonation: sand dunes in Lancashire.
This photograph from a landmark 1927 work by the
British biologist Charles Elton, *Animal Ecology*, shows
marram grass on the upper parts, dwarf willow on the
lower. Marsh is on the edge of the pond, and aquatic
plants under the water. Elton defined ecology as "scientific natural history" and stressed the study of populations rather than individual organisms.

144. Peat moor: English Lake District. Here, another
picture from Elton, we see the zones on the edge of a
tarn (*left to right*): floating leaved plants (water lily),
"reedswamp" of sedges (carex) mixed with floating
pondweed, marsh with carex grading into sweet gale
marsh, and bracken or coniferous woods. Study of the
distribution of plants and animals within communities
led to evaluation of succession within those communities and to analysis of the dynamic interactions characterizing them.

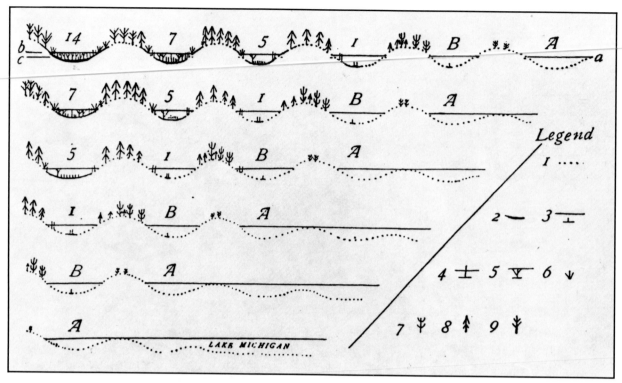

145. Pond stages. An early diagram by the American zoologist Victor Ernest Shelford, this outlines the development of new ponds near the shore of Lake Michigan and the aging of others. The vertical series at the extreme left is the history of the "present" Pond 14. The horizontal series at the top shows the present ponds, with intermediate types omitted. Shelford considered bioecology the "sociology of organisms."

146. Eltonian pyramid: Panamanian rain forest fauna. Charles Elton introduced the "pyramid of numbers" to show the quantitative relationships between producers and consumers in the food web (the network of food chains in an ecological community). As one moves up a food chain in a community, organisms tend to become fewer and bigger. A large number of plants may support a smaller number of herbivores, which may in turn be fed upon by still fewer carnivores.

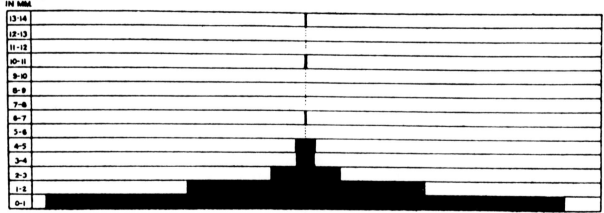

147. Competitive exclusion. In the 1930s the Soviet biologist G. F. Gauze (Gause) studied competition among yeasts and protozoans. He helped develop the principle that two species cannot occupy the same ecological niche at the same time. In an experiment reflected in these graphs, Gauze cultivated two species of paramecia separately and together with limited food resources. In the mixed population, the *Paramecium aurelia* eventually displaced the *P. caudatum*.

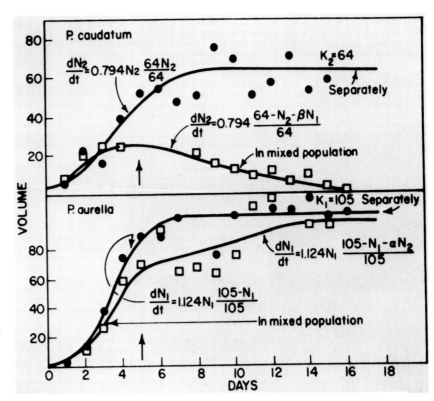

$$\frac{dN_2}{dt} = 0.794 N_2 \frac{64 N_2}{64}$$

$$\frac{dN_2}{dt} = 0.794 \frac{64 - N_2 - \beta N_1}{64}$$

$K_2 = 64$ Separately

In mixed population

$$\frac{dN_1}{dt} = 1.124 N_1 \frac{105 - N_1 - \alpha N_2}{105}$$

$K_1 = 105$ Separately

$$\frac{dN_1}{dt} = 1.124 N_1 \frac{105 - N_1}{105}$$

In mixed population

148. Carbon cycle. By tracing the movement of elements through the living and nonliving world, ecologists clarify the dependence of members of a community on primary resources like carbon, oxygen, nitrogen, other inorganic nutrients, and water. In this diagram of the carbon cycle for land areas, the widths of the pathways convey the relative contribution of each process to the whole.

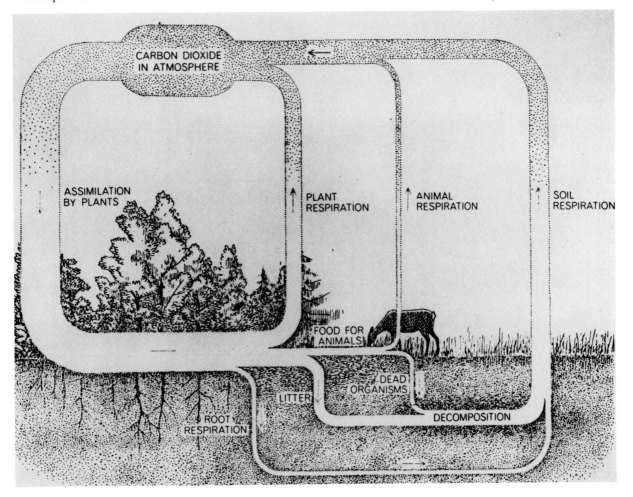

CARBON DIOXIDE IN ATMOSPHERE

ASSIMILATION BY PLANTS

PLANT RESPIRATION

ANIMAL RESPIRATION

SOIL RESPIRATION

FOOD FOR ANIMALS

DEAD ORGANISMS

LITTER

DECOMPOSITION

ROOT RESPIRATION

149. Overgrazed range. The place is New Mexico, the time 1953. The scene reminds us that the relationship of organisms to the environment has important economic implications. Indeed, ecological studies were initially tied very closely to agriculture. In the 1960s ecology became identified with popular environmental concerns, as the problems of pesticides, pollutants, and the preservation of natural environments emerged as public issues.

150. Problems of human intervention. Improper logging destroyed the vegetation on this Oregon hillside. The land could have been saved by selective harvesting and replanting of logged areas. Such management techniques depend on understanding the mutual relationships that sustain a given community of organisms—concepts developed in the first half of the century.

151. Infertile land. Ecological analysis showed that unproductive land could sometimes be made productive by correcting nutritional deficiencies. In 1942 the land shown here—in Australia's Southern Tablelands—was found to be lacking in a key trace element: molybdenum, which helps bacteria "fix" the nitrogen in the atmosphere, putting it into a form usable by other organisms.

152. Reclamation with fertilizers. Here we see what happened after the land was sown to clover and treated with fertilizers. It remained infertile.

153. Reclamation with fertilizers and molybdenum. This picture shows the same area of the Tablelands, but this time treated with tiny amounts of molybdenum as well as fertilizers. The clover flourished.

154. Artificially enhanced root systems. After mid-century, researchers found they could augment the root systems of plants by inoculation with soil fungi known as mycorrhizae. The fungi colonized the plant roots and, in effect, extended the root system. Pine seedlings inoculated with mycorrhizae have a dense root system (*bottom*), providing considerable root-surface area for absorption of water and nutrients. Such seedlings grow more rapidly, and thus have a greater probability of survival, than uninoculated seedlings (*top*).

155. Forest cutting and water yield.
How do forest-cutting practices affect water yields? The Rocky Mountain Forest and Range Experiment Station began a long-term study of this question in 1956. The researchers found that snow accumulated in the open areas and that the water yield was higher from the cutover areas than from the uncut areas.

156. Man-made forest. Kaingaroa State Forest in New Zealand was planted before 1940 on 300,000 acres of open land. By the time the photograph was taken about forty years later, Kaingaroa ranked as one of the largest man-made forests in the world. The principal type of tree planted was the Monterey pine, which had grown naturally only in California.

157. Living fossil. Thought extinct for nearly 70 million years, the coelacanth became a living presence in 1938 when a specimen was caught off South Africa. The primitive bony fish has lobe-shaped fins, structurally related to the limbs of higher vertebrates, and is itself related to the fishes from which land animals evolved. At left is the 1938 specimen with Marjorie Courtenay-Latimer, a museum curator responsible for saving the fish for study. The South African ichthyologist J. L. B. Smith named the fish *Latimeria chalumnae* in honor of her and of the discovery site—off the mouth of the Chalumna River. A number of specimens were subsequently caught off the Comoro Islands, the first in 1952. Below we see Smith with his hands resting on the 1952 specimen.

11

Evolution

Fossil evidence for evolution continued to accumulate during the century. Paleontologists established new links in the fossil record and began studying microscopic and soft-bodied forms preserved in sediments from the Precambrian era, the earliest period of geologic history.

Evolution as a field of research shifted gears in the middle decades, moving from the study and evaluation of fossils per se to analysis of the ongoing mechanism, or process, of evolution. Researchers now drew on a variety of approaches. Data from embryology, genetics, and ecology were supplemented by new insights yielded by the electron microscope and biochemistry. In addition, analysis of experimental demonstrations of evolution at work produced the basis for a comprehensive reformulation of evolutionary theory and brought together several complementary but distinct insights.

Early in the century much debate focused on how variation arose in individuals and how it accumulated in populations. The neo-Lamarckians, followers of the nineteenth-century French biologist Lamarck, said that changes acquired by an organism during its lifetime could be inherited. The neo-Darwinians argued that natural selection operated on small changes randomly occurring in a population. Based on his observations of animal populations in the Galápagos Islands, Darwin himself had proposed that evolution proceeded primarily through the accumulation of small inherited individual differences by members of a breeding population. Both genetic and ecological data were therefore relevant to the vigorous discussion in the early twentieth century. With the rise of genetics based on breeding studies, the neo-Darwinian point of view produced a new synthetic theory of evolution. Soviet science, however, was ruled until the 1960s by a kind of Lamarckianism, espoused by the biologist and agronomist Trofim Denisovich Lysenko.

The process of adaptive radiation, whereby one species evolves into several others through adaptation to different environments, was apparently responsible for the variety of species of finches observed by Darwin in the Galápagos. In the twentieth century adaptive radiation remained a fruitful field of investigation, studied in detail with the methods of genetics and ecology. Both geneticists and ecologists eventually showed that adaptive changes could be observed in populations responding to changes in the environment.

A case in point is so-called industrial melanism, the increased frequency of dark-colored (melanic) mutants in moth populations. In England the peppered moth (*Biston betularia*) was once gray. Around the mid-nineteenth century a few black or dark individuals (dubbed *carbonaria*) were noticed in Manchester, but they were easily seen by bird predators and could not readily survive. With extensive industrialization in many areas, tree trunks and rocks (often covered with light-colored lichens) became blackened with soot, rendering the pale form of the peppered moth extremely conspicuous. By the end of the nineteenth century nearly all the moths were dark. In the 1950s H. B. D. Kettlewell carried out field experiments indicating that the ascendancy of the *carbonaria* form was due to selective feeding by birds influenced by the moths' coloration.

Another important focus for twentieth-century research has been the role played in species formation by isolation of populations. Here genetic and evolutionary theory developed together. Using mathematical analysis of genetic variation—and focusing on populations that had been isolated by environmental, physiological, or behavioral factors—researchers studied the response of populations to the introduction of new genes and the redistribution and spread of such genes in later generations. Comparison of theoretical analyses with actual data from the field showed that marked physical and physiological variation often occurred within a given species. Investigation of such microevolution—minor change over short periods of time—revealed the inheritance of groups of genes that corresponded to specific habitats.

The synthetic theory of evolution that emerged in the second quarter of the century considered how populations adapt genetically to changing environments. It linked genetic change to mutation, genetic recombination, and migration of individuals into the gene pool (that is, to genetics proper), as well as to environmental determinants permitting the survival of certain variants; together, all these factors lead to evolutionary change over time. Correlation of these results with the new structural and biochemical data available after mid-century prompted reconsideration of the common features shared by lower forms of life. This yielded new insight into organisms' evolutionary relationships, and biologists revised the traditional division into kingdoms. Instead of two kingdoms, there were now five, based on level of development (single cell with or without a nucleus, multicellular) and mode of nutrition (photosynthetic, absorptive, ingestive).

Thus, twentieth-century evolutionary theory gradually came to include not only analysis of fossil evidence of primitive forms but also study of the subcellular structural features of microorganisms, analysis of the origin of organelles, and evaluation of the biochemical capabilities of the simplest forms of life. The rise of molecular biology, fostering a new understanding of the molecular basis of inheritance, provided added incentive to develop a comprehensive theory explaining the origins and reproduction of life. These issues are surveyed in Chapters 16 and 17.

159. The pulse of life. An American history of the earth published after Morgan's book contained this summary of evolution through geologic time (not shown to scale). The numbers on the bottom graph indicate the first records then known of, for example, vertebrates (1), dinosaurs (5), and man (11).

158. Evolution of the elephant's trunk. The geneticist T. H. Morgan used these pictures in a 1916 book explaining the theory of evolution. Commenting on how the presence of large numbers of an organism facilitates evolutionary change, he wrote: "When elephants had trunks less than a foot long, the chance of getting trunks more than one foot long was in proportion to the length of trunks already present and to the number of individuals; but increment in trunk length is no more likely to occur from an animal having a trunk more than one foot long than from an animal with a shorter trunk." At left is *Macritherium* (*top*) and *Tetrabelodon* (*bottom*); above is an African elephant.

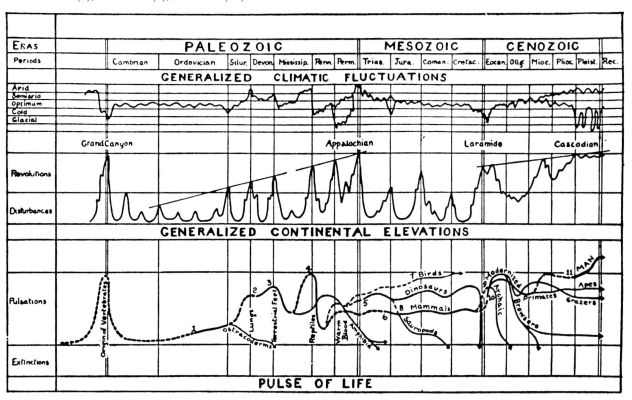

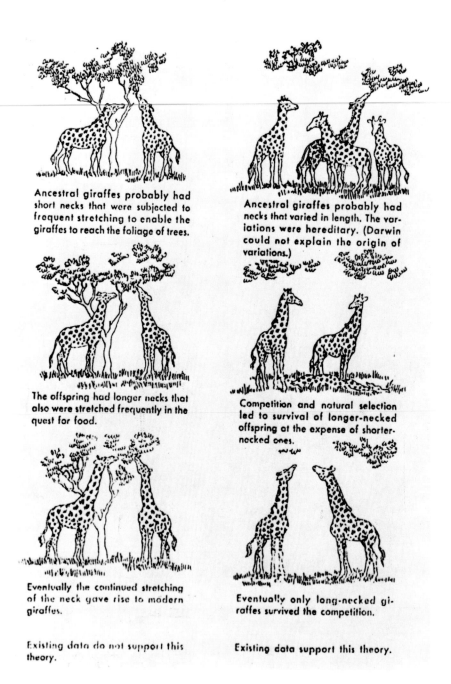

Ancestral giraffes probably had short necks that were subjected to frequent stretching to enable the giraffes to reach the foliage of trees.

Ancestral giraffes probably had necks that varied in length. The variations were hereditary. (Darwin could not explain the origin of variations.)

The offspring had longer necks that also were stretched frequently in the quest for food.

Competition and natural selection led to survival of longer-necked offspring at the expense of shorter-necked ones.

Eventually the continued stretching of the neck gave rise to modern giraffes.

Eventually only long-necked giraffes survived the competition.

Existing data do not support this theory.

Existing data support this theory.

160. Lamarck versus Darwin. A major theme in much of nineteenth-century evolutionary thought was the notion, espoused by Lamarck, that characteristics acquired by organisms during their lifetime could be inherited by their offspring. In contrast, Darwin held that slight variations among offspring would give some animals advantages over others in competing for limited resources and these advantageous traits would be inherited. Darwin called this process natural selection, comparing it to the artificial selection practiced by breeders. Darwin did not totally reject the inheritance of acquired characteristics; he thought that environmental conditions might affect the germ plasm but that hereditary differences among offspring were more important to natural selection. Above is how the Lamarck-Darwin conflict was explained in an American textbook published after the mid-twentieth century.

161. Adaptive radiation: Hawaiian honeycreepers. Adaptive radiation is the evolution of a single species into a number of different species adapted to particular features of their environment. The various Hawaiian honeycreepers in this drawing are all descendants of a single species that came to the islands a few million years ago. The drawing is from the classic 1947 work by the British zoologist David Lack *Darwin's Finches*. Evidence of adaptive radiation of finches and other creatures on the Galápagos Islands provided Darwin with data for his theory of evolution by means of natural selection. Analysis of species formation has been a major theme of twentieth-century biology.

162. Rapid evolutionary change. England's peppered moths were once pale gray. But as industrialization blackened tree trunks and rocks in many areas with soot, the moth's pale form became extremely conspicuous. By the end of the nineteenth century almost all the moths were dark. In the picture a dark moth is held by a robin, and a dark form and a regular form rest on the bark.

163. Peppered moth: new and old. The left two specimens are from the mid-twentieth century. The right two specimens were caught eighty or ninety years earlier, when the moth's black coloration had not yet completely developed. Careful field experiments in the 1950's by H. B. D. Kettlewell showed that the ascendancy of the dark form was a result of selective feeding by birds influenced by the moths' coloration.

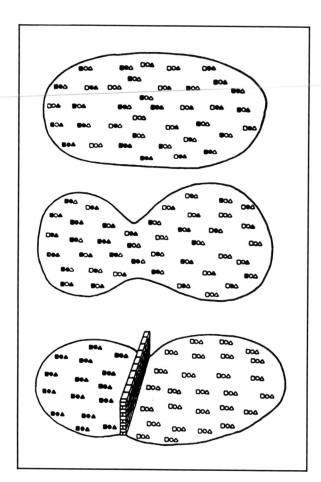

164. Isolation and species formation. The role that isolation can play in the origin of species is illustrated in these diagrams by the Russian-born American geneticist Theodosius Dobzhansky. At top, the gene pool of a population includes some differences (exemplified by a black variation and a white variation), but intermixing occurs freely within this one species. If different variations come to predominate in different areas, the species divides into "races" (middle picture). The bottom picture shows the result of the isolation (indicated by a wall) of one area from the other; each of the two populations is distinct and genetically unmixed, with hybridization now impossible.

165. Microevolution: the varied shells of the European land snail. The common European snail *Cepaea nemoralis* is highly polymorphic; that is, many genetically different classes exist. The shell may be yellow, pink, or brown, with from zero to five bands. (The bands are counted on the shell's biggest whorl.) Research at mid-century showed that the shell variation bears a marked relation to the "microhabitat" of the snail; in other words, certain shell patterns are associated with specific physiological characteristics. The scale in the picture shows millimeters.

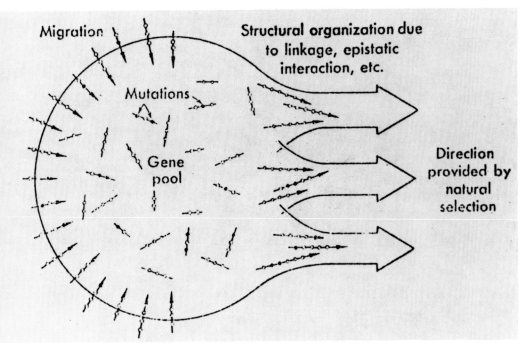

166. Synthetic theory of evolution. How do the genetic characteristics of a population of organisms change in response to a changing environment? Biologists came to attribute evolutionary change to four interacting processes: mutation, genetic recombination (from intercrossing between members of the population and between members and new individuals migrating into the population), structural changes in chromosomes, and natural selection.

167. The five-kingdom classification. In the second half of the twentieth century most biologists gave up trying to classify organisms into just two kingdoms: plant and animal. At lower evolutionary levels, it was not clear what characteristics might distinguish plant from animal. A five-kingdom classification gained broad currency. The one in this diagram was proposed in 1969. The kingdom of the Monera (bacteria and blue-green alga) consists of primitive organisms whose cells are termed procaryotic—that is, they lack a well-defined nucleus surrounded by a membrane. The Protista (including protozoans, certain algae, and some funguslike forms) are unicellular organisms. The Plantae, Fungi, and Animalia are multicellular. This system embodies three levels of development. At each level organisms diverge in their mode of nutrition. In the Monera photosynthesis and absorption are found, and in the two higher levels ingestion as well.

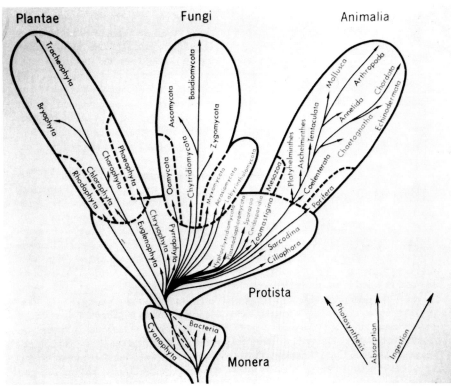

168. Fossilized bacteria and algae.
These procaryote remains were found in the Gunflint Chert of Ontario, Canada, a formation dated at about 1.9 billion years in age. Such microfossils from the Precambrian era provide evidence for the earliest stages of evolution. Fossil evidence of this kind was unknown before mid-century.

169. Jellyfish from Ediacara. After World War II the Australian geologist Reginald Sprigg discovered fossils of soft-bodied marine animals in the Ediacara Hills north of Adelaide. The fossils dated from the Precambrian, which apparently experienced a much more complex level of evolution than scientists had supposed. Among the animals represented in the Ediacaran fauna were jellyfish, sea pens, worms, and previously unknown forms. Similar fossils of about the same age were subsequently found elsewhere. The fossil shown in the picture is of the jellyfish *Cyclomedusa davidi*. The scale is in centimeters.

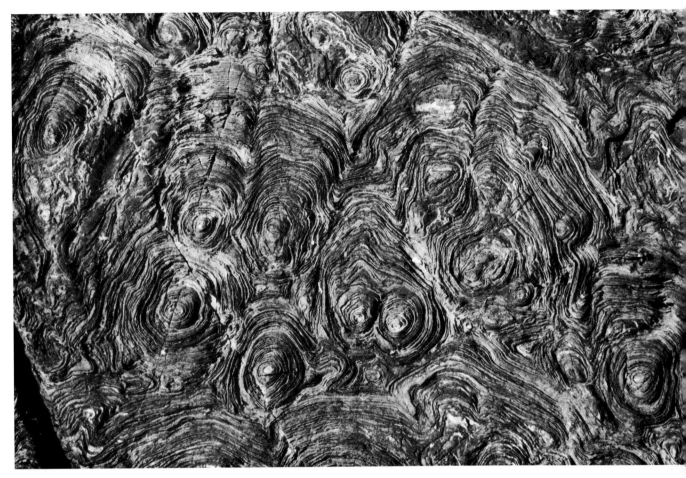

170. Remains of Precambrian algae. Much older than the Ediacaran fossils are these stromatolites—bunlike masses of limestone deposited by blue-green algae—from the Great Slave Lake in Canada. They are dated at about 2 billion years. Above, we see the layered internal structure of the stromatolites in a horizontal cut through several masses; below is a view of the surface structure.

Part Four

INVESTIGATING
LIFE
PROCESSES

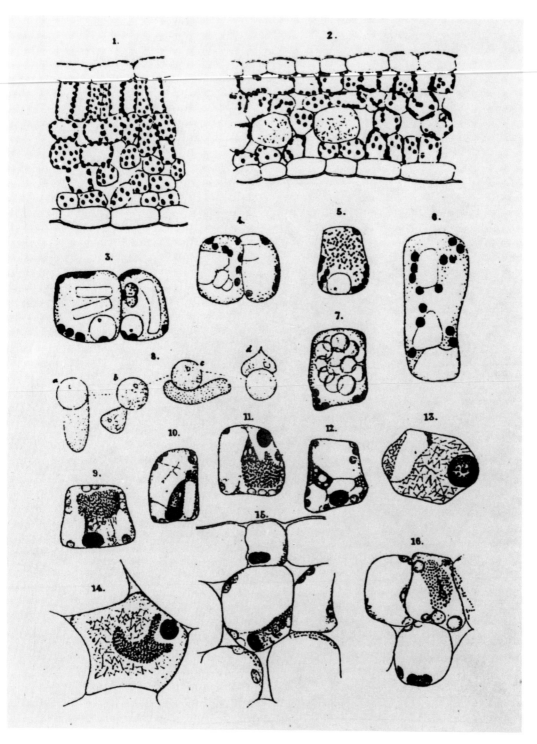

171. Plant cells infected with tobacco mosaic virus. These optical-microscope observations, published in 1903 by the Russian botanist D. I. Ivanovsky, reveal virus crystals and "inclusion bodies" in plant cells. Ivanovsky mistakenly believed tobacco mosaic disease was caused by bacteria too small to be retained by standard filtration methods. He had in fact discovered viruses, organisms scientists would not begin to understand until the introduction of electron microscopy.

12

Structure

At the beginning of the twentieth century biology was still primarily an observational science. Researchers' most important tool was, without question, the microscope. Thanks to it the cell theory, which asserted that organisms consist of basic units called cells, had become firmly entrenched.

Drawings from the early part of the century show to what a large extent biological knowledge rested on careful observational methods. Staining techniques introduced in the 1870s had made visible not only individual parts of the cell but also bacteria, the agents of disease predicted by Pasteur's germ theory. Tissues to be studied were fixed in a preservative, embedded in paraffin so that they could be thinly sliced, and finally stained to reveal specific structures. Although biological researchers were resorting more and more to experimentation, anatomical and histological investigation—the largely descriptive study of structure—would remain basic research emphases throughout the twentieth century. Eventually, when linked together with chemical studies, they extended scientists' insight into biological structure down to the level of the molecule.

The cell's internal organization and features revealed by the optical microscope had impressed researchers ever since the cell theory was put forward in the nineteenth century. Because of vastly improved optics, magnification, and staining techniques, however, twentieth-century biologists were able to appreciate far more fully than their predecessors the functional significance of different parts of the cell. Observation under very high magnification or with the oil-immersion lens revealed fine detail that sometimes could be seen to change over time or to vary under different experimental conditions. Scientists began to note that certain structures and subcellular events were associated.

Cell division, a favorite subject of the latter nineteenth century, continued to attract much interest in the twentieth. The movement of the nuclear bodies called chromosomes and the partitioning of a cell's contents into two new daughter cells were thoroughly studied. Photography and microcinematography captured these processes more fully than simple observation had. They allowed a closer look at fine detail and eliminated some of the element of subjective interpretation found in scientists' drawings. Study of the movement of chromosomes was especially important to research on genetics and development in the early decades of the century.

The invention of the electron microscope in the 1930s brought a revolution in cell research. The new instrument opened up to observation very fine subcellular structure that had previously been invisible to scientists, and the cell emerged to view as an extraordinarily complex system. After mid-century, using the new techniques available for separation and chemical analysis, biologists isolated some of the newly revealed subcellular particles and studied their individual properties. They found that photosynthetic processes in plants, for example, are localized in the complex internal structure of the subcellular unit called the chloroplast. Researchers also analyzed ribosomes, where proteins are synthesized. The electron microscope showed larger units, such as the hairlike cilia and flagella associated with locomotion in one-celled organisms, to have a much more complex structure than had previously been recognized. For cilia and flagella this structure was found to be uniform in all living organisms. Thus, the electron microscope made cell ultrastructure an important focus for research, and cell biology emerged as a discipline.

The transmission electron microscope that unveiled this wealth of fine structure was analogous to the optical microscope, but with electrons (rather than photons) passing through the thin specimen. New visualization techniques were soon developed that displayed surface phenomena. The use of shadowing, for example, and the scanning electron microscope revealed exquisite structures on specimens as diverse as a fungus spore and a growing plant tip. Appearing in popular magazines highlighting the wonders of science, micrographs made with these techniques captured the public's imagination with their beauty. This chapter's photographic images of a spore, the lung of a bird, and the head of a fly remind one of the exquisite detail of the hand-drawn plates in Robert Hooke's *Micrographia*, which appeared in 1665. By the third quarter of the twentieth century photomicrography was creating visual images even of macromolecules.

When integrated with biochemical techniques, observational methods revealed much about the internal functioning of the cell and about the biology of physiological processes like the formation of a blood clot and the immune system's response to foreign bodies. Viruses had not been visible under the optical microscope, but the electron microscope showed them to have a precise external structure and a complex life cycle. Fundamental processes like the replication of genetic material or the contraction of muscle revealed an unexpected beauty and symmetry, raising new questions about the nature of the molecular events underlying basic biological processes. The development of radioactive tracing techniques for following biochemical processes eventually made it possible to explore these questions. While observation of structure suggested how parts might function, only such experimental techniques could show how the events actually occurred.

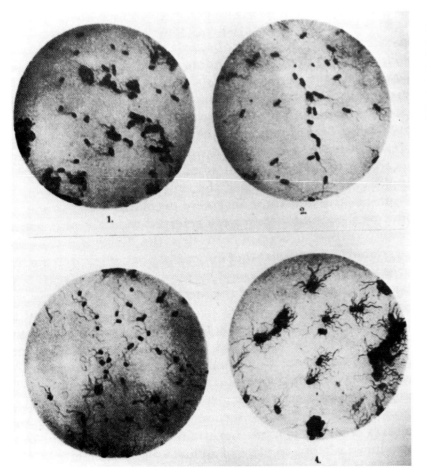

172. Bacteria, with flagella. By the turn of the century biologists had discovered most of the external features of bacterial cells. The German microbiologist Friedrich Loeffler took these photomicrographs in 1890. They are apparently the first ever made showing the taillike flagella of bacteria.

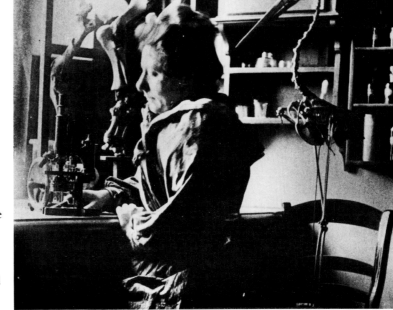

173. Perfecting observational skills. Like many of her compatriots, the American physiologist Ida Hyde studied abroad. In 1896 she became the first woman to receive a doctorate at the University of Heidelberg in Germany. Here she is seen with the tools of her trade, her microscope and zoological specimens. In 1902, Hyde became the first woman elected a member of the American Physiological Society.

174. Specimens on slides. Optical microscopy has remained a vital tool for biologists in the twentieth century. Histology, which studies the structure of tissues, is a case in point. Shown here is a histology laboratory, where slides of specimens are made and studied. Preserved tissues are cut thin, and the resulting "sections" are then stained so that their structure can be viewed. Staining jars (or boxes) sit on the table in front of the histologist, who is examining sections from the tray on her right. The photograph, from the American Museum of Natural History, was taken about 1930.

175. First pictures by microcinematography of the division of a cell. The development of successful tissue culture methods in the early part of the century made it possible to study activities of live tissues, such as movement and division. In time, biologists added cinematography to their selection of microscope research techniques. The pictures reproduced here were taken by the pathologist R. G. Canti in Britain in the early 1930s. The clock in the upper right-hand corner gives an idea of how long it takes for the cell to divide.

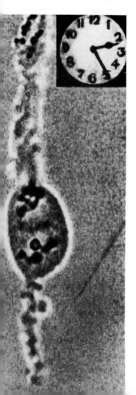
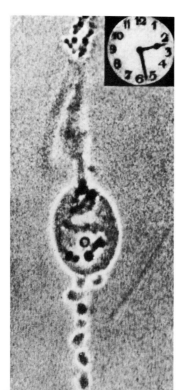
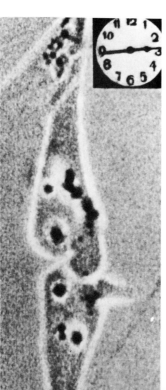
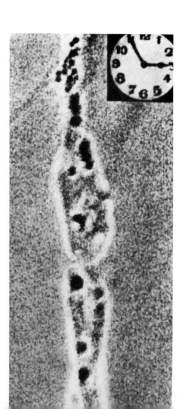

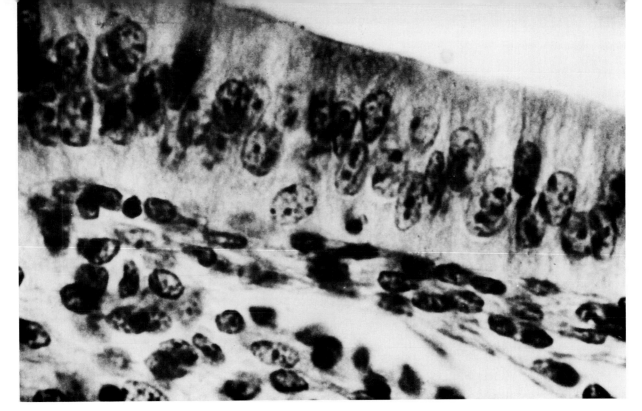

176. Rat uterus cells. Here the optical microscope was used to monitor histological changes during an experiment. The photomicrograph, from the 1960s, shows how the uterus of the rat responds to the female hormone estrogen. The top layer of cells is much thicker than usual. Accompanying biochemical studies showed that this effect was due in part to increased protein synthesis in the cells.

177. A typical cell. This schematic diagram from a textbook of the early 1970s shows the complex structure of the cell as revealed by electron microscopy. Note how much more internal detail is evident here than in the photomicrograph above. The nucleus and the cytoplasm outside it are filled with minute structures associated with specific processes like cell respiration, protein synthesis, and cell replication. The biochemical events involved in these processes began to be identified with specific structural components of the cell in the latter half of the century, as researchers used autoradiography, fractionation (techniques for isolating cell parts), and chemical analysis to unravel the complex events occurring in each living cell.

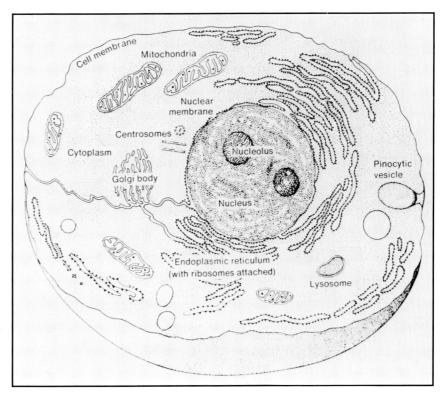

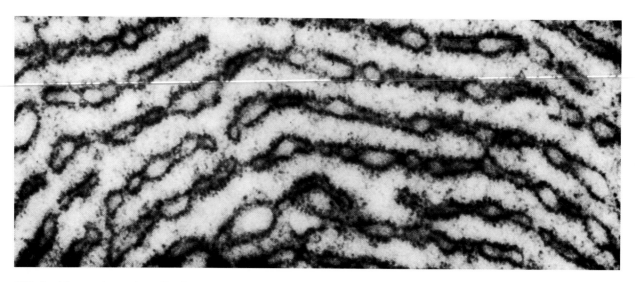

178. Inside a guinea pig cell. Electron micrographs like this one showed that the cytoplasm, which under the optical microscope had seemed to be largely undifferentiated substance, was in fact extremely complex in structure. It contained a network of tiny tubes—the endoplasmic reticulum—and large numbers of tiny particles. These so-called microsomal particles, later termed ribosomes, proved to be the site of protein synthesis in the cytoplasm. The micrograph was taken about 1955 by George Palade, a cytologist at the Rockefeller Institute in New York.

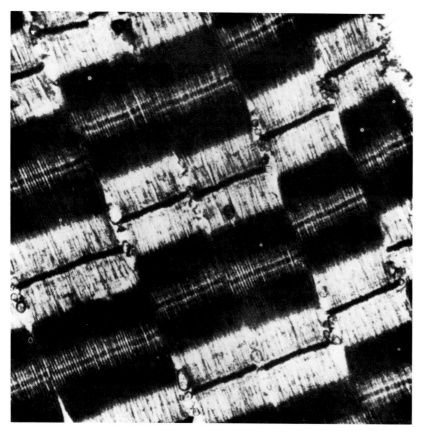

179. Striated muscle from a rabbit. The muscle fibrils shown here were magnified 24,000 diameters. The characteristic banding, or striations, of voluntary muscle had been visible under the optical microscope. The physiological significance of each of these zones, however, became clear only in the 1950s, when scientists began to correlate the properties of the contractile proteins actin and myosin with the structure and behavior of the bands. Such analysis resulted in the formulation of the sliding filament theory of muscle contraction.

180. Animal and plant. Below left is a cross section of cilia from the gill of a freshwater mussel; PM indicates the plasma membrane enclosing the fibrils. Below right is a longitudinal section of part of the flagellum (a whip-like appendage used for locomotion) of a green single-celled alga. Visible parts include peripheral fibers (PF), central fibers (CF), and the basal body (BB) within the cell. Flagella and cilia are structurally and functionally similar, although flagella are much longer and generally less numerous. The magnification is greater at left than at right.

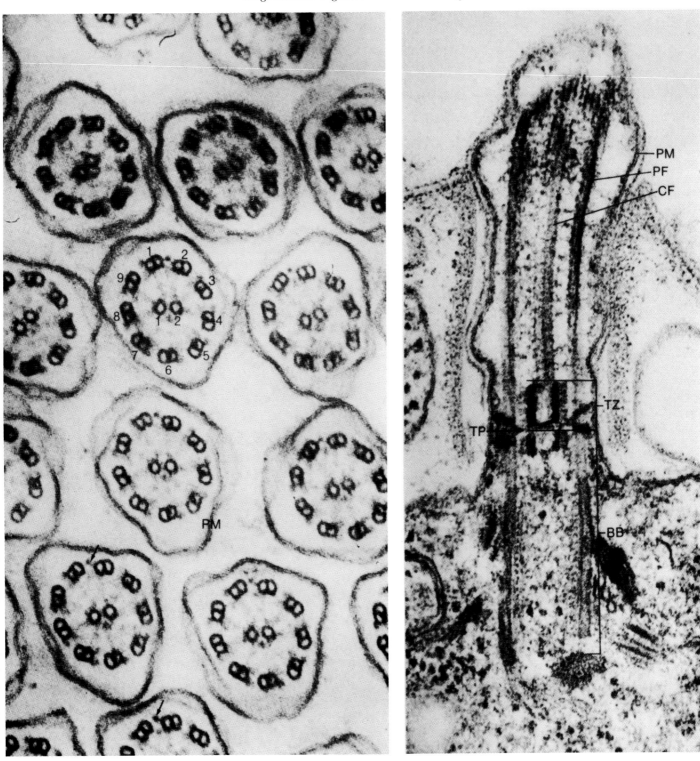

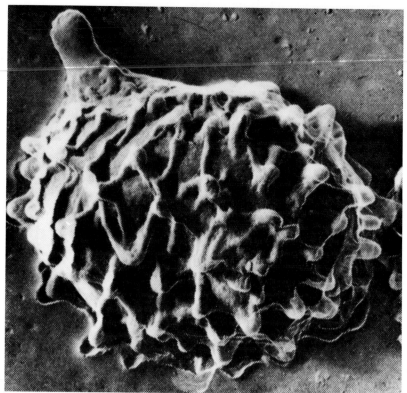

181. Fungus spore: shadowed carbon replica. The internal structure of a specimen can be studied with the electron microscope by taking a thin slice, or section. To view a specimen's surface, a replica and shadowing may be used. This replica is from a spore of the fungus *Russula mairei*. It was made, about 1950, by depositing a film of carbon on the spore. Shadow casting—for the production of a three-dimensional effect—was achieved by depositing a thin layer of material at an angle to the replica. The investigation of spore and pollen structure has been particularly important in the study of asthma.

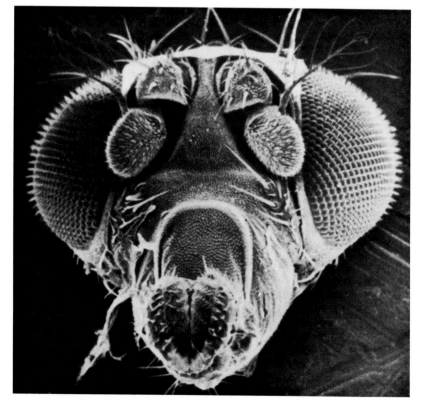

182. *Drosophila*. Here one of the major players in twentieth-century biology, the fruit fly, confronts us. A large compound eye bulges out on each side of the fly's head, which possesses sensory hairs, some on the mouth parts (*bottom*). The clarity of the photomicrograph reveals the intricate geometry of minute sensory structures. Study of the function of these organs has revealed features common to sensory mechanisms in many types of living things, features not necessarily evident from structural studies alone.

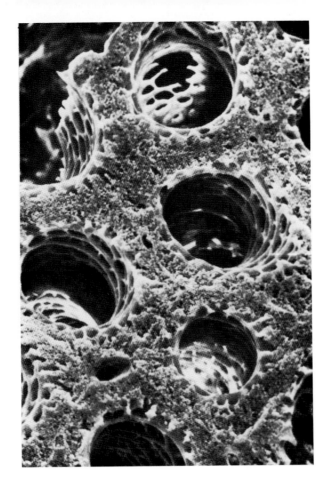

183. Bird lung. The cylindrical tubes prominent in this cross section from the chick of a domestic fowl are actually less than 0.5 millimeter in diameter. They are called parabronchi. In the 1960s analysis of the flow pattern and respiratory exchange of these airways revealed how distinctive anatomical features of the bird lung affect its physiological capabilities. While mammal lungs have saclike alveoli, the bird lung tubules are open at both ends, permitting continuous unidirectional flow of air and the consequent extraction of more oxygen.

184. Apex of a celery plant. This is where new leaves begin. Each of the ridgelike structures around the dome of the apex is a young leaf. The leaves emerge at equal intervals if the temperature is constant. The micrograph was made with a scanning electron microscope, which sweeps across the specimen with a thin beam of electrons. Unlike the conventional transmission electron microscope, where electrons pass through the specimen, the scanning electron microscope needs no replica. The specimen was shadowed with gold. By studying changes in the plant's surface, scientists seek to discover how cellular and macroscopic changes are correlated in the normal processes of growth.

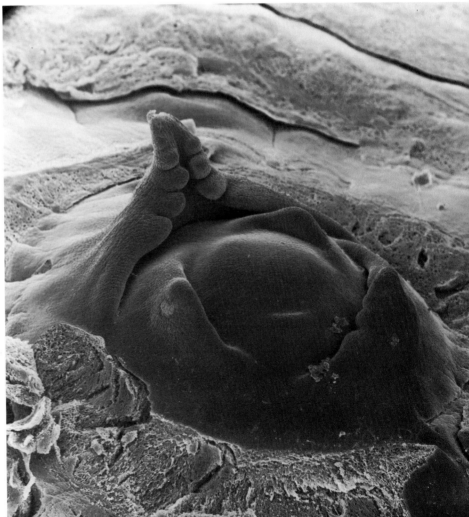

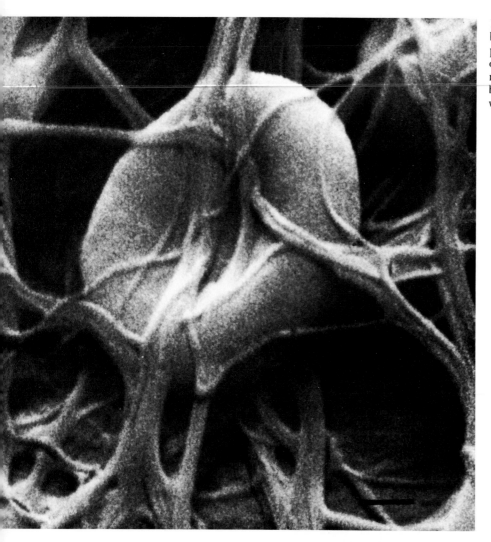

185. Red blood cell. This human red blood cell is enmeshed in fibrin, the protein that makes up the main body of a blood clot. Red blood cells of mammals generally have, as here, a biconcave shape—like a doughnut with an incomplete hole.

186. White blood cell. Here is the type of white blood cell called a macrophage. Much larger than the red blood cell, it serves the body's immune system by gobbling up, or phagocytizing, debris and particles like bacteria. The electron micrograph was made in the mid-1980s with the help of a computer-assisted technique. The cell was still alive; it was not dried or coated with a metal film.

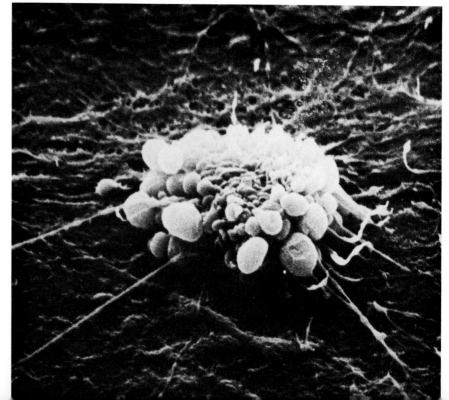

187. Tobacco mosaic virus. Long after Ivanovsky's pioneering efforts, the virus that causes tobacco mosaic disease continued to be closely studied. It was the first virus to be isolated in pure form. In this 1960s electron micrograph, the chromium-shadowed viruses show clearly the rodlike form common to plant viruses. The tiny balls are polystyrene calibration spheres (PSL), measuring 880 angstroms (88 billionths of a meter). They give an idea of the viruses' size.

188. Plasmid. A plasmid is a ring-shaped molecule of DNA that can replicate itself. The type shown here was used as a carrier molecule, or "vector," in the early 1970s in the first pioneering experiments in genetic engineering. It is about 3 microns in circumference and carries a gene conferring resistance to the antibiotic tetracycline. The plasmid was cut, spliced to another plasmid, and introduced into a bacterial cell, where it replicated and expressed genetic properties new to that cell.

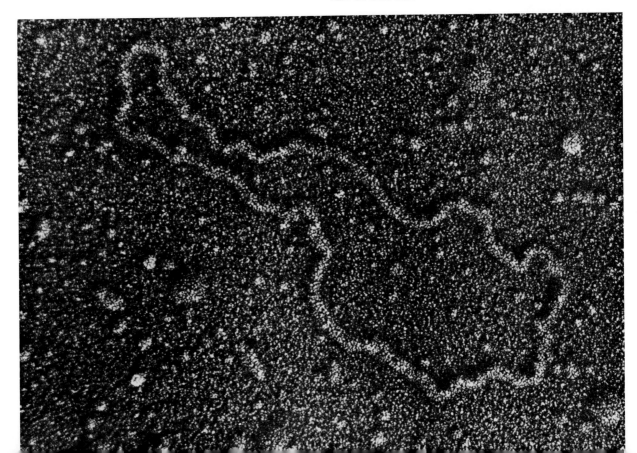

189. Atwater-Rosa respiration calorimeter. In the late nineteenth century the American chemist and physiologist Wilbur Olin Atwater, with the help of the physicist E. B. Rosa, devised a successful respiration calorimeter for research on human metabolism and nutrition. Based on previous German work, the instrument measured the heat given off by the subject, confined in the windowed enclosure at left.

190. Testing food needs of a child. By the early 1920s respiration calorimetry and related devices had come into widespread use. The equipment at the left measured the amount of oxygen consumed and carbon dioxide produced by the child (in the respiration chamber at the right) while breathing. To assess further the physiology of heat production, the pulse rate was also counted and recorded.

13

*F*unction

While the analysis of structure was traditionally the main avenue by which scientists studied how organisms function, twentieth-century researchers increasingly developed distinctive experimental techniques to examine the processes of life. They sought to adapt the methods of physics and chemistry to the study of living matter, but the unique properties of living organisms made this task difficult. As a result, complicated life processes were first dissected into their component parts with an array of new tools. Physiologists, for example, used specialized mechanical and electrical apparatus that could isolate and monitor specific functions like respiration or the transmission of nerve impulses. Studies of structure were thus regularly supplemented and enhanced by functional research that noted change in a given phenomenon over time. Gradually, direct analysis of change and the mechanisms of that change replaced the study of structure as the primary focus of investigation.

The new apparatus that supported such experimental studies became ubiquitous in biological research laboratories early in the century and a recognized feature of hospital care in later decades. As instruments like the revolving drum recorder and the oscilloscope challenged and occasionally replaced the microscope as the primary tool in biological research and in medical diagnosis, they created another kind of visual image emblematic of living matter: the graph, a visual representation of functional change.

Research instruments from the turn of the century show scientists' growing confidence and pride in the newly refined techniques of measurement that had become popular in the latter decades of the 1800s. Scientists confined human and animal subjects alike in experimental spaces from which respiratory gases, blood samples, and urine and other excreta were drawn and analyzed. They used special chambers called calorimeters to study metabolism and nutrition, and they measured and chemically analyzed the gases emitted in order to determine precisely organisms' needs under given environmental conditions. Such experimental apparatus, derived from a variety of nineteenth-century prototypes, was often connected to recording machines that produced a graphic record.

In the early decades of the twentieth century experimental systems of this kind were used extensively to study function in both animals and human beings. During the years between World War I and World War II, researchers adapted physiological techniques to the study of a variety of problems, analyzing, for example, the physiology of work for industrial applications. Such techniques gradually evolved into numerous procedures for analyzing the effects of exercise, work, and fatigue and for evaluating the effects of underwater and high-altitude conditions on human physiology—research that eventually contributed to the exploration of both sea and space. At the same time, examination of differing respiratory and nutrition requirements among animals heightened appreciation of the variety and complexity of life-forms and contributed to detailed analysis of physiological control mechanisms. In the process, human and comparative physiology emerged as distinct fields of research.

Widespread adoption of recording techniques was central to the experimental analysis of function. Turn-of-the-century investigators relied on registration devices first used by physiologists in the mid-1800s. The "kymograph," or recording drum, produced a "tracing," or graph of changing function. Tracings on sooted paper recorded events too rapid to be observed with the naked eye. Preserved in this way, physiological processes were measured and evaluated with ease after the experiment ceased. These visual records produced by mechanical registration techniques were gradually modified as researchers adopted electrical and electronic methods of detection and transmission. With their strip-chart recordings of electrical events in the heart and the brain, the electrocardiograph and electroencephalograph are descendants of this family of mechanical research instruments. The electrocardiograph and electroencephalograph became important diagnostic tools in medicine in the second quarter of the century.

After mid-century, monitoring instruments were incorporated into routine hospital procedures. Surgery and intensive care became associated in popular experience with these tools of physiologists. Police departments, too, used physiological instrumentation, in the form of the polygraph machine. The first modern lie detector, built in 1921 by John A. Larson, a medical student at the University of California, simultaneously recorded blood pressure, pulse, and respiration.

By the middle decades the line of the graph was not the only symbol in the public mind for changing function. The moving beam of the oscilloscope also became a well-recognized popular image. The graph's wavy line and the oscilloscope beam conveyed to the viewer the state of vital processes, distinguishing life from death and signaling the existence of ordinarily invisible and undetectable physiological events.

Records from the various instruments in use during the century all revealed the continuously changing states of hidden internal processes. Such data suggested to scientists that complex control mechanisms operated within the body, coordinating a diverse set of functions. The concepts of homeostasis (an internal constancy or equilibrium) and biological feedback were developed in the century's second quarter as biologists began to appreciate these dynamic processes and the intricate regulatory mechanisms operating within living organisms.

Full elaboration of such concepts, however, awaited the development of specialized chemical techniques capable of interrupting physiological processes at crucial points and identifying transitory, intermediate states, captured only in the ascending or descending curve in the graphic image. While the graphic record allowed visualization, measurement, and analysis of physiological function over time, it could not by itself provide the data to understand the molecular events presumed to underlie these physiological phenomena. Chemical analysis would provide another powerful set of symbols with which to construct an understanding of life. Eventually, these chemical images would dominate the language biologists used to describe life processes. As we shall see in Chapter 14, research on integrative mechanisms in the first half of the century illustrates very clearly the complex dialogue emerging from the synthesis of new data being assembled by anatomists, physiologists, and biological chemists.

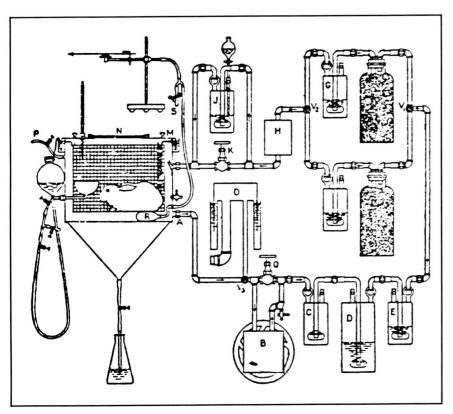

191. Respiration apparatus for a small animal. This type of research apparatus was perfected by Francis Benedict, who worked under Atwater for a short period before the latter's death in 1907. The system is a closed circuit. It uses the same air, after purification, again and again. Air leaves the chamber at **A** and is measured by the spirometer **O**. **B** is a blower that keeps the current in motion. Bottles **D** and **E**, containing sulfuric acid, remove the moisture. Carbon dioxide is removed in bottle **F**, which contains soda lime. The glass vessel **J** contains water, which provides moisture for the animal's comfort.

192. Laboratory research on human subjects. Arlie Bock's laboratory at the Massachusetts General Hospital was just a few years old when this 1926 photograph was taken. One of the first physiological research laboratories to be established in a general hospital, it was a precursor to the Harvard Fatigue Laboratory (1927-1947), the first laboratory for comprehensive study of the normal human being. In the photograph are (*left to right*) Bock, David Bruce Dill (later the research director of the Harvard Fatigue Laboratory), and Lewis Hurxthal.

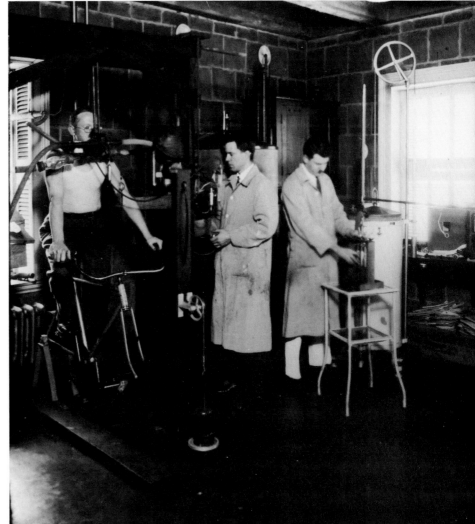

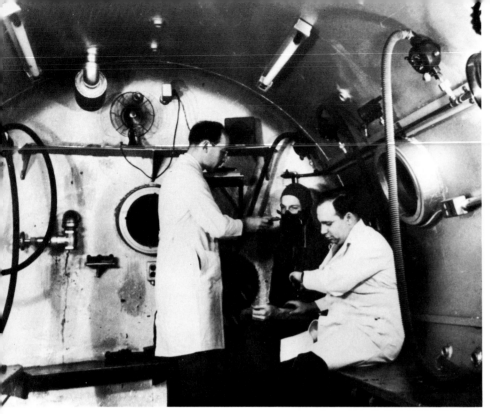

193. Decompression chamber. During World War II the extraordinary physical stresses experienced by pilots and crew of high-performance aircraft became a matter of serious concern. This decompression chamber, photographed near the end of the war, was located at the Ohio University aviation physiology laboratory. The physiologist Fred Hitchock used it in research on explosive decompression and high altitudes.

194. Human centrifuge. World War II aviators faced the possibility of blackouts and unconsciousness during high-acceleration combat or dive-bombing maneuvers. Shown undergoing its initial trial in 1942 at the Mayo Foundation is the first modern U.S. human centrifuge. Research with devices like this helped develop ways for pilots (and later, astronauts) to cope with high g forces and made significant contributions to medical theory and practice regarding, for example, the cardiovascular system.

195. The effects of exertion. This specially trained parakeet was photographed during flight in a wind tunnel. The plastic face mask and attached tube collect all the air exhaled by the bird. Researchers in the 1960s found that parakeets' oxygen consumption was lowest at a certain flight speed (35 kilometers per hour) and that flying slower or faster than that speed required the consumption of more oxygen.

196. Measuring respiration: an indirect approach. Plethysmography—the determination of variations in the volume of an organ or limb as the amount of blood in it changes—dates back to the nineteenth century. In this body plethysmograph, the baby's face is outside the chamber, the rest of the body inside; the volume displaced within the chamber as the baby breathes reveals the amount of air entering and leaving the lungs.

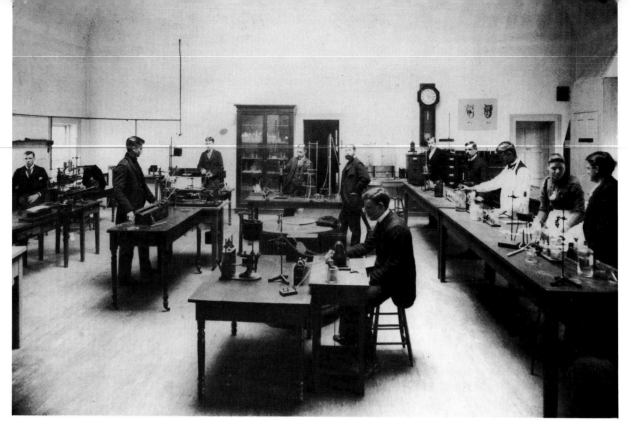

197. Student lab facilities: turn of the century. Much
of the equipment in this University of Michigan physiology laboratory was obtained in Germany at the end
of the nineteenth century. The students in the lab are
gaining experience in mechanical, electrical, and chemical procedures. At center and at rear left are kymographs, recording drums producing the graphic
records now so familiar in physiological research.

198. Harvard physiology lab. In a
1905 scene, dental students pose with
American-made drum kymographs
for recording experimental results.
These kymographs, unlike the individually crafted instruments imported from Germany, were manufactured locally in quantity with
interchangeable parts. As instrumentation became increasingly integral to
physiological research, scientific instrument manufacture became an
important industry. Note the electric
lights, a relatively recent innovation
in this laboratory.

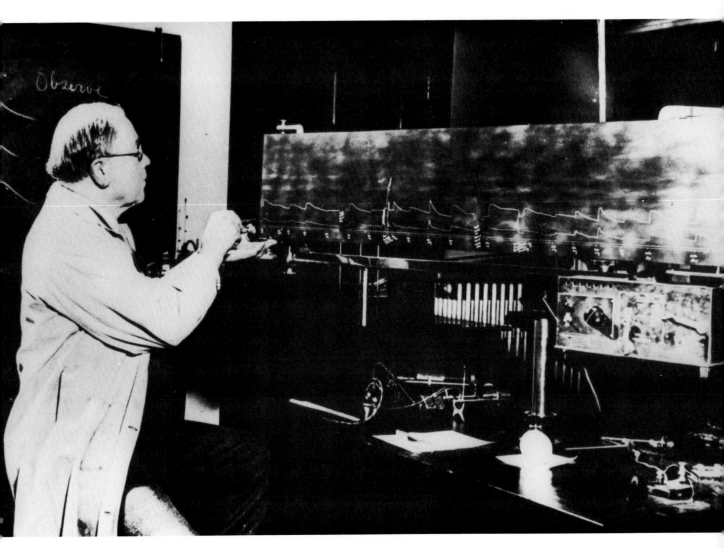

199. Cannon at the kymograph. In the first decades of the century the nervous system was a major focus of research for the American physiologist Walter Bradford Cannon. This 1940 photograph captures Cannon in his Harvard laboratory investigating neuromuscular transmission. The tracings made on the long paper kymograph show the effect of drugs on nerve action. For nearly a century the kymograph, driven by a clockwork mechanism or electricity, was the major research tool in physiology and pharmacology. Larger drums and long rolls of paper allowed physiologists to record multiple events simultaneously over extended periods of time.

200. Lie detector. This long paper kymograph was used by the San Diego Police Department. Monitoring with such apparatus the respiration, blood pressure, and other physiological processes of a suspected criminal during questioning is based on an observed linkage between mental and physiological states. Nonetheless, lie detectors are not absolutely reliable indicators of the telling of truth or falsehoods.

201. Recording heart activity. At the beginning of the century the Dutch physiologist Willem Einthoven invented a practical electrocardiograph to record electrical activity in the heart. Among the variations that were quickly developed was this electrocardiograph with immersion electrodes. It was built in 1912 for the renowned British cardiologist Sir Thomas Lewis.

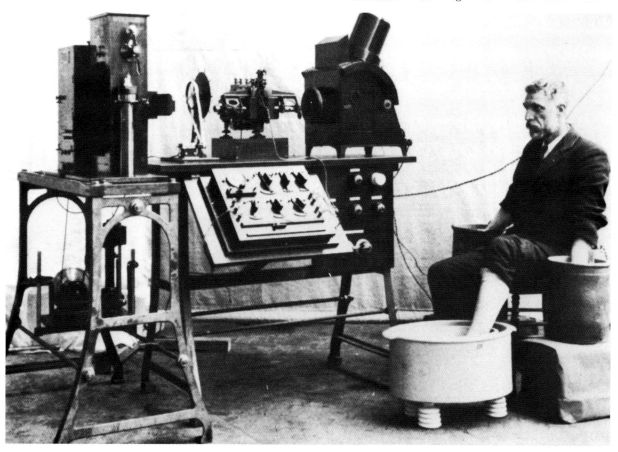

202. String galvanometers. In the early decades of the century the string galvanometer, perfected by Einthoven, was the principal device used to detect bioelectric currents. Current from the body was passed through a delicate wire suspended between the poles of an electromagnet, which caused the "string" to move in proportion to the current; the string's image could be recorded on photographic film. This room, at Washington University, also has a cathode-ray oscilloscope, widely used from the 1930s to study electrical currents in nerves and muscle.

203. Electroencephalograph, 1930s. Hans Berger, a German psychiatrist, in the 1920s made the first recording of the human brain's electrical activity with electrodes placed on the skull. The EEG gained importance during World War II when it was used in diagnosing brain dysfunction resulting from injuries.

204. Using electroencephalography in research. With improvements in electronics and the introduction of computer-enhancement techniques, electroencephalography was applied to increasingly sophisticated research. Here, a London scientist investigates links between brain waves and intelligence. The electrical activity of the girl's brain is monitored as she listens to a series of tones. Note the graphic displays to the left and right of the girl, and the charted graphs at the rear.

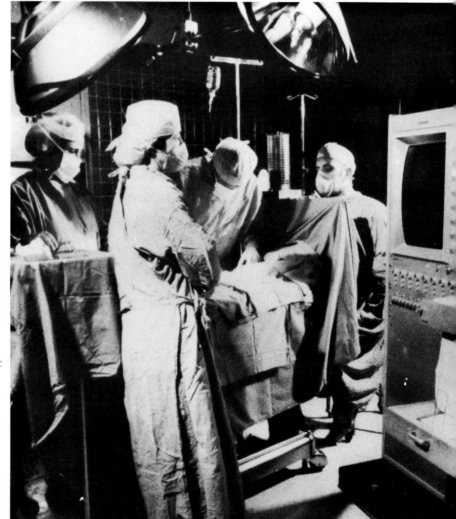

205. Monitoring during surgery. An array of imaging and other electronic monitoring devices became available to surgeons in the latter decades of the century, as laboratory techniques gained a place in medical practice.

206. Bedside monitoring. Systems available to hospitals by the 1970s included electronic bedside monitors providing direct display of the heartbeat count and electrocardiogram waves. A central nurses' station might have a large oscilloscope screen displaying the electrocardiograms of several patients at the same time.

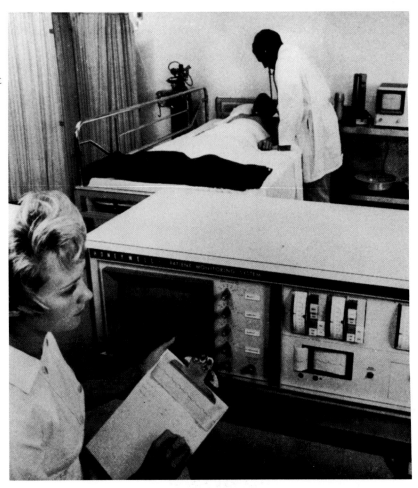

207. Telemetering from space. When U.S. astronauts went into space, their physiological responses were transmitted back to earth. The physiological and other relevant data were summarized on "data sheets"—one for each consecutive interval of time—that were later evaluated by physicians studying the effects of spaceflight on the human body. Shown here is part of a typical data sheet for astronaut L. Gordon Cooper, from his Project Mercury orbital mission in 1963. The time interval covered runs from ten seconds after liftoff to twenty seconds after.

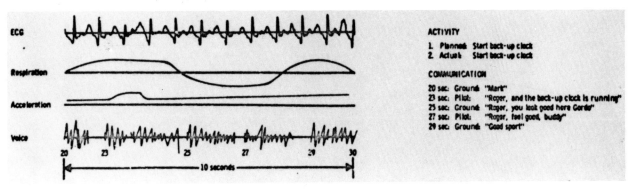

208. Pavlov and the conditioned reflex. The concept of the conditioned, or learned, reflex was developed by the Russian physiologist Ivan Pavlovich Pavlov in a series of classic experiments beginning at the turn of the century. By ringing a bell before giving food to a dog, he trained the dog to salivate at the sound of the bell. Pavlov's research on the physiology of the brain and higher nervous activity grew out of his work on the physiology of digestive secretion. This earlier work—which earned him the Nobel Prize in physiology or medicine in 1904—was based on classic vivisectional techniques of nerve section and nerve stimulation. By linking explanations of nervous and mental function, Pavlov profoundly influenced the development of both physiology and psychology in the twentieth century.

CHAPTER

14

Integration

When nineteenth-century scientists wanted to study individual physiological processes in detail, they separated the processes and analyzed them in isolation. By the early twentieth century, however, there was a growing appreciation of how physiological processes are integrated with one another in the living organism. Physiologists were starting to link together, to synthesize, the results of disparate studies and to reevaluate underlying relationships and assumptions.

Particularly in neurophysiological research it was apparent that new methods and techniques would be required in order to study how whole organisms function. Perhaps the most famous innovator of this period was the Russian physiologist Ivan Pavlov. His work on conditioned reflexes played a central role in both physiology and psychology in the twentieth century.

Early research instruments and techniques were gradually modified and adapted for studying events in the whole organism. The myograph, or "muscle writer," for example, was originally designed in the 1850s to record events in the isolated muscle and nerve. By the beginning of the twentieth century researchers routinely used it to study more complex phenomena like muscle fatigue or the effect of the mind on the knee-jerk reflex. Such research involved whole organisms rather than isolated organs or tissues.

Physiologists and psychologists saw that complex events like muscle fatigue and the knee-jerk reflex were coordinated in the whole animal

through the integrated action of the nervous system. Early in the century this integration was studied most notably by the British physiologist Charles Sherrington. Using the traditional vivisectional techniques of cutting or stimulating a nerve, together with kymographic recording techniques, Sherrington mapped neural pathways throughout the whole body.

The existence of another important coordinating mechanism was suspected at the turn of the century, fostering a new line of investigation. This work depended on the refinement of analytical techniques derived from chemistry. In the 1890s scientists had found that extracts from certain tissues acted like drugs in distant parts of the body. This led to the formulation in 1905 of a new biological concept, the hormone. Hormones are discrete chemicals released in one part of the body that stimulate activity in another part. After isolating one such substance, "secretin," which stimulates secretion by the pancreas, the British physiologists William Bayliss and Ernest Starling suggested that there were additional hormones responsible for coordinating the orderly sequence of events in the digestive tract. A full-scale search ensued for such substances.

The French physiologist C.-E. Brown-Séquard had postulated in 1889 that extracts of the testes and ovaries had a rejuvenating effect. Secretions from these glands were suspected to be responsible for sexual differentiation and sexual function. However, because of their distinctive chemistry, their low concentration, and their association with the controversial subject of human sexuality, these hormones proved elusive—at least until other research made the subject more scientifically appealing. In the 1910s and 1920s biologists began studying in detail the role of the sex glands in the development of the organism. Initially, they relied on anatomical evidence visible upon dissection or microscopic examination to suggest the results of hormone action. Important work by the American embryologist Frank Rattray Lillie on the freemartin in twins—a freemartin is a genetic female modified apparently by the influence of hormones from its male twin—provided further evidence suggesting the existence of male and female hormones. Research on sex hormones gained momentum and prestige in the 1930s and 1940s, as one by one these potent substances were isolated and characterized.

Hormone research applied not only the analytical methods of the anatomist and physiologist but, increasingly, those of the biological chemist. This brought scientists with different laboratory skills into collaboration, opening the way to new discoveries and the isolation of specific hormones. The earliest purified hormones were used to treat particular diseases. Isolation of the hormones of the thyroid gland and the pancreas in the 1910s and 1920s led to treatments for cretinism and diabetes. Such successes fed a growing expectation among life scientists that potent chemical substances like these would be found in other tissues as well and that they were likely to be important in the coordination of a multitude of perhaps yet unrecognized biological processes. The substances' presence was suspected in every tissue. Discoveries—some immediate, some not—were many. Plant hormones, or auxins, were found responsible for plants' responses to light. Lower organisms, too, were found to possess a variety of characteristic chemical mediators that controlled development, molting, and even complex behavior patterns.

The discovery of neurotransmitters—chemicals that help in the transmission of nerve impulses—showed that the relationship between neural and chemical coordination was complex and inadequately understood. It became clearer and clearer that the connection between stimulus and response was not necessarily a simple one. Rather, a combination of neural and chemical signals together gave rise to a complex sequence of events producing a given physiological phenomenon or a given behavior sequence.

The isolation and characterization of specific hormones, so striking a feature of the work of the second quarter of the century, was followed after mid-century by detailed analysis of the mechanism of hormone action. This work focused first at the subcellular and later at the molecular level. At the same time, scientists extended research on neural pathways as the chemical characteristics of specific parts of the nerve cell were determined—for example, the terminal vesicles at the synapse. Evidence from both electron microscopy and biochemical research played a role in most of these later physiological studies.

Anatomical, physiological, and chemical studies were thus pursued together as the experimental ethos permeated twentieth-century biology, transforming its content and direction and shifting its focus first to the whole organism and then to the properties and behavior of specific molecules acting within the organism. The holistic emphases of the first several decades receded as, at mid-century, reductionist techniques drew attention to individual molecules within the cell, creating new economic opportunities based on further biological research.

209. Neuromuscular research. These beginning-of-the-century University of Michigan students are studying complex physiological processes. The laboratory equipment is based on apparatus that Michigan Professor Warren Plimpton Lombard used in Europe in the 1880s in the laboratories of Carl Ludwig in Leipzig and Angelo Mosso in Turin. At top, the student fatigues the muscle of his left hand. His responses are recorded on the kymograph at left. In the middle picture, two students work together —one as subject, the other giving an electrical stimulus to motor points on the first student's hand. At bottom, a shrouded student's response (the knee-jerk reflex) to an external stimulus is recorded on a kymograph drum. Each of these experiments involves the whole person rather than an isolated organ. The physical responses measured, therefore, are influenced by the mental state of the experimental subject. (The shroud in the bottom picture is an attempt to limit this influence by minimizing distraction.) Physiologists and psychologists alike made use of such quantitative experimental methods.

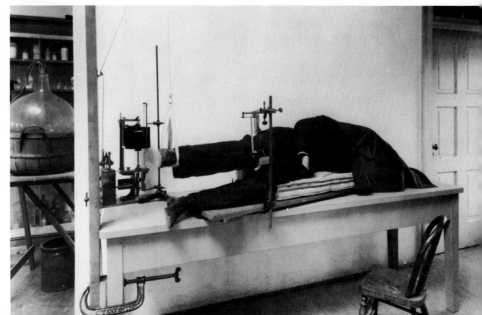

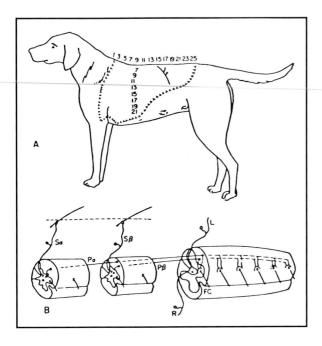

210. Sherrington on the scratch reflex. The British neurophysiologist Charles Scott Sherrington established a holistic, integrative view of nervous function in higher animals. He found that reflex actions involve not only activation of one set of muscles but "inhibition," or relaxation, of opposing ones. His 1906 monograph *The Integrative Action of the Nervous System* is a classic of twentieth-century science. In the diagram, the "receptive field" is outlined at top (**A**). The scratch reflex of the left hind leg can be evoked by stimulating this saddle-shaped area. Below (**B**), the spinal "reflex arcs" are diagrammed. **L** and **R** are receptive nerve paths from the left and right feet. **S** indicates paths from hairs on the dog's left side. **FC** is the final, common path—the motor neuron that activates the hip muscle.

211. Sensory and motor homunculi. By mid-century scientists were able to map or represent schematically the relative areas of brain tissue devoted to the various parts of the body. The type of diagram reproduced here was developed by the brain surgeon Wilder Penfield. It shows for one half, or hemisphere, of the cerebral cortex the location and relative size of (1) the areas in the "sensory strip" (*left*) receiving sensory information from different body parts and (2) the areas in the "motor strip" (*right*) controlling specific body areas.

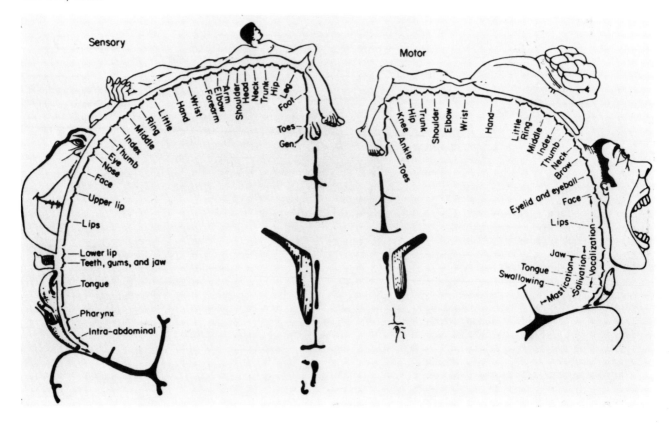

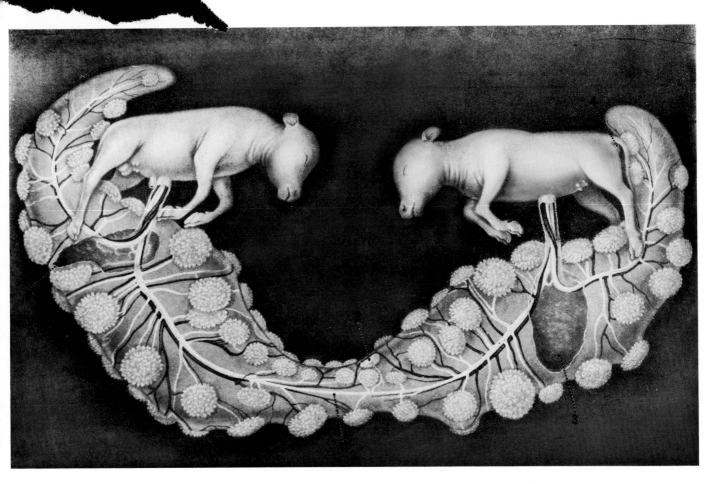

212. Sex hormones and the freemartin. The American embryologist Frank Rattray Lillie was partly responsible for focusing biologists' attention on questions of the origin, nature, and action of sex hormones. In a major 1917 paper, from which this illustration is taken, he gave a thorough analysis of freemartinism in cattle. Bovine twins may be two females, two males, or a male and a freemartin, an imperfectly developed and usually sterile female. In bovine twin pregnancies, the two placentas often fuse, uniting the blood circulations of the fetuses. As the freemartin twin develops, its reproductive system takes on male characteristics under the influence, according to Lillie, of male hormones from the male twin. The illustration shows the chorion (or extraembryonic) membrane of a cow with twins, male (*left*) and freemartin (*right*).

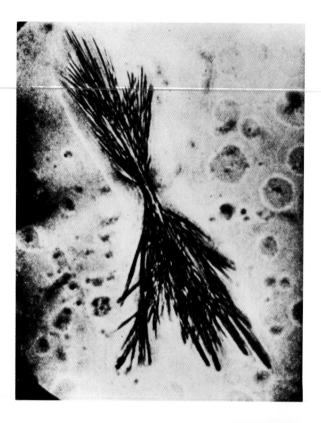

213. Thyroxine crystallized. The early decades of the century saw marked progress in the isolation and study of the secretions of the endocrine glands. The active component of thyroid hormone, thyroxine, was first isolated in crystalline form on Christmas Day 1914, by the American biochemist Edward Kendall. Kendall was able to produce 35 grams of the hormone from 3 tons of fresh cattle thyroid. Research on the physiological effects of such potent substances showed that complex internal processes were regulated by chemical as well as nervous stimuli. The study of hormones required the combined talents of anatomists, physiologists, chemists, and physicians.

214. Thyroxine: before and after treatment. Thyroxine, like its therapeutic predecessor, thyroid extract, was used to treat such consequences of thyroid deficiency as cretinism. The child shown here was deficient in thyroid secretion and stood just 37 inches tall at age ten (*left*). Given thyroxine for a year, she grew 6 inches (*right*). The great British physician Sir William Osler exclaimed, "Not the magic wand of Prospero, nor the brave kiss of the daughter of Hippocrates, ever effected such a change as that we are now enabled to make." The marked physical changes induced by thyroxine convinced skeptics of the reality of this hormone. The physiological effects of other hormones were more elusive, generating much heated debate in the first two decades of the century.

215. Genesis of the discovery of insulin. The isolation of insulin was perhaps the greatest success in endocrinology in the early decades of the century. This page, dated October 31, 1920, is from the notebook of the Canadian surgeon Frederick Banting. Scientists had suspected that the so-called islets of Langerhans in the pancreas produced a hormone that could prevent diabetes by enabling the body to metabolize sugar properly, but attempts at treating diabetes through oral administration of fresh pancreas or pancreatic extracts had been unsuccessful. Banting surmised that the active substance in the pancreas might be destroyed during extraction by digestive enzymes produced in the pancreatic cells known as acini. He thus proposed tying off, or ligating, the pancreatic ducts and waiting for the acini to degenerate before trying to isolate the putative curative substance. (The islets of Langerhans degenerate much more slowly.) Although Banting's analysis was not entirely correct, it was the basis for a research program that led in 1921 to the isolation of insulin and to a successful treatment for the dreaded "sugar disease," diabetes.

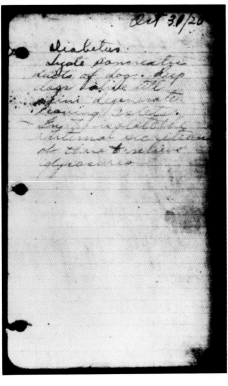

216. Insulin pioneers. This 1921 photo shows Banting (*right*) with his graduate-student assistant Charles Best and one of the diabetic dogs they used to prove the effectiveness of insulin. The insulin work was carried out in the Toronto University laboratory of physiologist J. J. R. Macleod. The biochemist James Bertram Collip, enlisted in the project by Macleod, was responsible for isolating insulin in a chemically pure form. The discovery of insulin stimulated the search for other hormones in the following decades. A more immediate result was that Banting and Macleod won the 1923 Nobel Prize in physiology or medicine. Banting gave half of his prize money to Best; Macleod gave half of his to Collip.

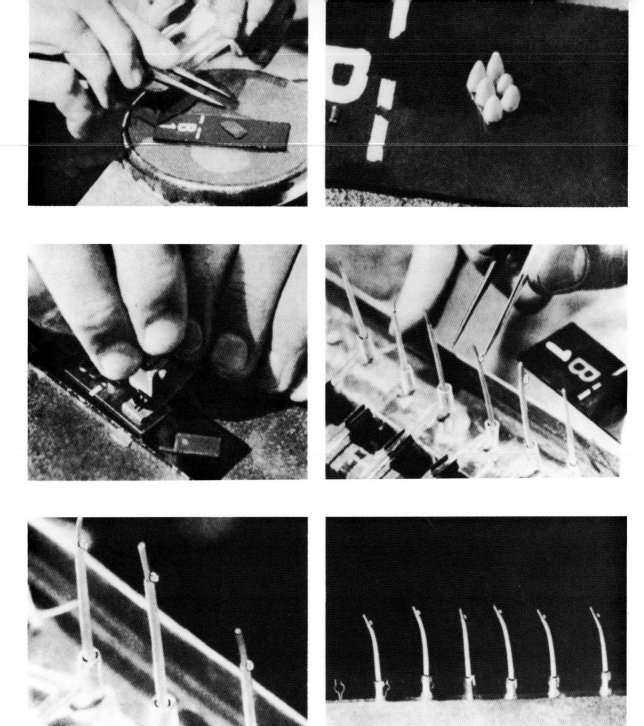

217. Plant hormones. The existence of plant hormones was demonstrated in this classic experiment by the Dutch botanist Frits Went in the 1920s. Many scientists suspected that plants' growth and responses to light and gravity were controlled by a special "growth substance." Went put tips of oat seedlings on agar (gelatine) so that the growth substance could diffuse into the agar. The agar was cut into tiny blocks and applied to one side of other growing seedlings. These grew faster on the treated side, and so they curved. Experiments such as these demonstrated the important role of chemical substances in the coordination of growth, as well as the ubiquity of hormones in plants and animals alike.

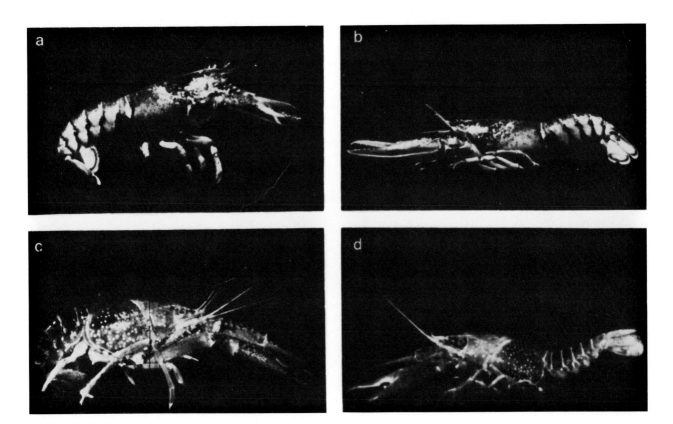

218. Neurohormones in action. The 1970s saw increased research on neurohormones in animals. In one experiment, lobsters (*above*) and crayfish (*below*) showed opposing postures when injected with two different neurohormones. Serotonin (*left*) activates flexor muscles, producing an aggressive posture, as if the animal were preparing to fight. Octopamine (*right*) activates extensor muscles, producing an extended body posture, as if the animal were going to flee. Such experiments illustrate how chemical and neural signals work together.

219. Reflex pathways. This table from the late 1960s summarizes the various possible ways in which a reflex action can be activated by a stimulus. Note the variety of combinations of nervous and chemical signals that can effect a response. At the beginning of the century only pathway B1 was recognized. The word "hormone" was not even coined until 1905. The table thus reflects the cumulative achievements of neuroendocrine science since then.

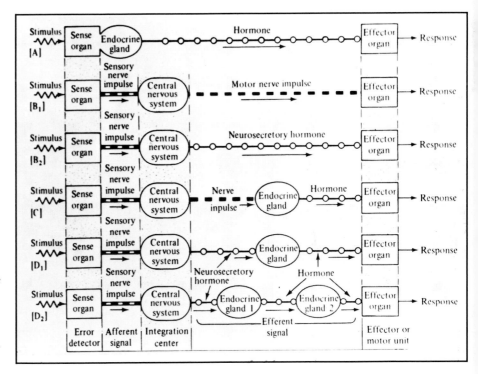

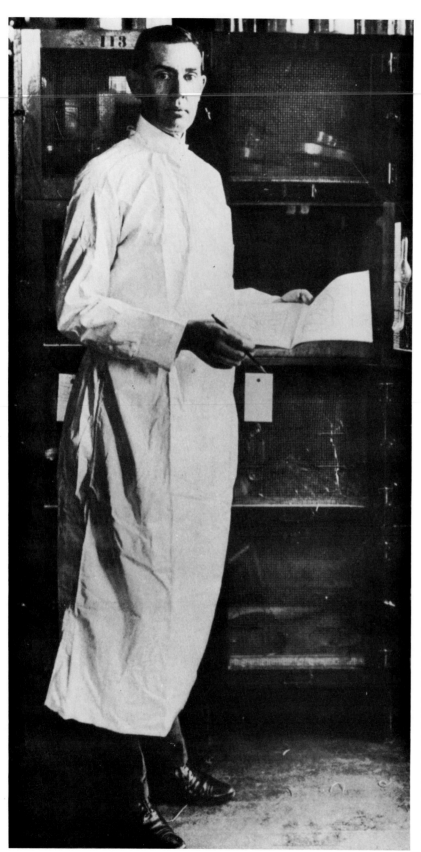

220. McCollum. The American biochemist Elmer Verner McCollum was a pioneer in the use of small animals for controlled experiments. Before World War I he established the first white rat colony in the United States for nutrition research. McCollum subsequently identified vitamins A and B and played a key role in the study of vitamin D. Research on how such discrete chemical substances affect the growth and reproduction of animals required the collaboration of anatomists, physiologists, and chemists. This encouraged the rapid growth of what was at the time called experimental biology.

15

*B*iochemistry

The twentieth century saw researchers wed the techniques of anatomical study and physiological analysis to those of chemistry, with remarkable results. Animal and plant substances—large, chemically complex molecules that were difficult to separate from one another—gradually yielded to analysis, shedding new light on many fundamental biological problems. In the nineteenth century chemists had for the most part eschewed biochemical research because biological substances appeared to be so complex. Toward the century's end, however, organic chemists made significant advances in deciphering the structural patterns of organic molecules. In the twentieth century the isolation and structural analysis of macromolecules like proteins represented the cutting edge of research.

By the turn of the century physiologists and microbiologists had already noted that distinctive chemical substances like enzymes and antitoxins play important roles in the body. Researchers saw that further chemical analysis of the products of living processes was necessary not only for continued exploration of the body's structure but also as the key to understanding how complex processes like growth and metabolism are coordinated. The important new concepts of hormone and vitamin were introduced in 1905 and 1912, respectively. Anatomical and physiological evidence for the existence of these classes of chemicals was rapidly supported by the actual isolation and characterization of several different hormones and

vitamins. Along with enzymes, antitoxins, and antibodies, these were the most important groups of substances with special biological properties to be studied in the first half of the twentieth century. Their successful characterization required their separation from other compounds found in cells— tasks initially dependent on anatomical and physiological assay techniques. It was hoped that the chemical characterization of these important molecules would eventually lead to their artificial synthesis and commercial marketing on a large scale.

Analysis of the biochemical events accompanying nutrition and growth required a ready supply of organisms for study. Experimenters thus selected for use those animals, plants, or microorganisms whose physiological and nutritional states could be regulated in the laboratory. Early in the century nutritional studies began making routine use of laboratory-bred mice and rats. Later, microorganisms like the intestinal bacillus *Escherichia coli* were cultivated and harvested for research. Also, cell culture techniques provided a convenient source of biological molecules for organic chemists to analyze and ultimately synthesize. All this work depended on the cooperation of specialized research personnel drawn from zoology, botany, physiology, microbiology, and chemistry. The work also necessitated a new cadre of technical workers who managed the animal colonies and maintained the specialized analytical equipment that was used.

The analysis of major structural components of cells, such as proteins and carbohydrates, preoccupied biochemists in the early decades of the century, as did the isolation of chemicals with special physiological properties like hormones, vitamins, and enzymes. These primarily analytical efforts were gradually superseded by study of the lengthy and complex biochemical pathways and feedback mechanisms that make up the sequences of reactions of, for example, intermediary metabolism in animals and photosynthetic processes in plants.

The biochemical techniques used in this research were strongly shaped by the German chemical tradition, which blossomed in the latter decades of the nineteenth century. In the early years of the new century leading researchers from around the world often trained in German-speaking countries and adapted techniques first developed in laboratories there. For example, Emil Fischer's work on the chemical structure of carbohydrates and proteins was fundamental to understanding the chemical characteristics and physiological features of living organisms.

Functional studies supplemented and greatly extended Fischer's analytical approach. Otto Warburg and, later, Hans Krebs used manometric techniques to monitor metabolic processes in the laboratory. Their methods for sampling biochemical events in progress in slices of tissue were eventually perfected with the general introduction of radioactive tracers in the 1940s. From mid-century, researchers widely adopted the spe-cialized techniques of ultracentrifugation, chromatography, and electrophoresis, spurring further study of the transformation of specific substances within organisms, tissues, and cells. These methods, too, were enhanced by monitoring the presence of radioactive tracers with Geiger or scintillation counters.

The new ability to sample biochemical events in progress revolutionized the nineteenth-century tradition of input-output studies based on analysis of foodstuffs, gases, urine, and feces. Twentieth-century biochemists studied both the structural basis of living organisms and how chemicals are transformed within them. The increasingly sophisticated analytical techniques eventually incorporated automated procedures, and laboratories were gradually industrialized, as biochemical research took on commercial importance, most notably in the production of pharmaceuticals and agriculturally important chemicals. Thus, the integration of several methods of analysis fostered the emergence of two institutional contexts, the research laboratory and the industrial laboratory.

It was in this setting that, about mid-century, experimental biology was transformed into molecular biology. This conceptual revolution reverberated in all areas of twentieth-century biology. It focused research onto the molecular level and, by creating the possibility of engineering biological events, catapulted the life sciences into new economically important arenas.

221. Wieland's laboratory. The organic chemist Heinrich Wieland was for many years a professor at the University of Munich. He received the Nobel Prize in chemistry in 1927 for his work on the structure of bile acids from the liver. The classical laboratory methods of organic chemistry included physical separation of a compound, analysis of its physical and chemical properties, and determination of its elemental composition. At the turn of the century, organic chemists developed new techniques for studying unknown compounds; they converted them, either by degradation or transformation, into substances of known composition.

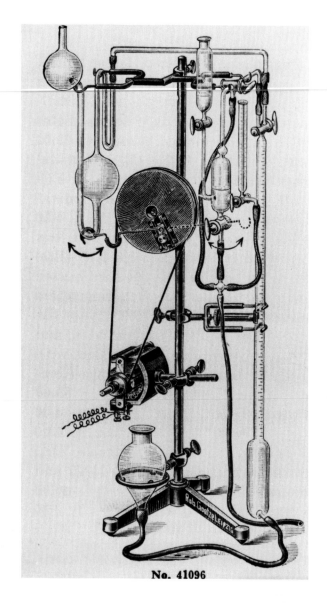

222. Van Slyke apparatus for determining amino groups. The American biochemist Donald Van Slyke was particularly noted for his skill at devising analytical equipment. The device shown here, in a drawing from a 1914 commercial apparatus catalog, was developed by Van Slyke in 1911. It was used to determine the number of free amino groups (NH_2) in a peptide or protein by measuring the amount of nitrogen gas released when the groups reacted with nitrous acid. "A complete determination of the NH_2 nitrogen of amino acids requires," said the catalog, "but six to ten minutes." The entire apparatus, excluding the motor, cost just twenty-five dollars.

No. 41096

223. Crystalline cortisone. One of the more important commercial products of the century's biochemical research, cortisone was extracted from ox bile and later prepared synthetically from plant products. It was one of many substances that were isolated and then became powerful drugs in the therapeutic armamentarium of medicine. In addition to cortisone and other hormones, these substances included vitamins and antibiotics. The numerous new "biologicals" expanded the pharmaceuticals market in the first half of the century.

224. Cortisone researchers. This picture was taken in a Mayo Clinic laboratory in 1950 after the announcement of the year's Nobel Prize in physiology or medicine, which was awarded in recognition of work on hormones of the adrenal cortex. Present are (*left to right*) physicians Charles H. Slocumb and Howard Polley, biochemist Edward Kendall, and physician Philip Hench. Kendall and Hench shared the Nobel with the Swiss chemist Tadeus Reichstein. Among the achievements honored were Kendall's isolation of cortisone and Hench's successful use of the hormone in treating rheumatoid arthritis.

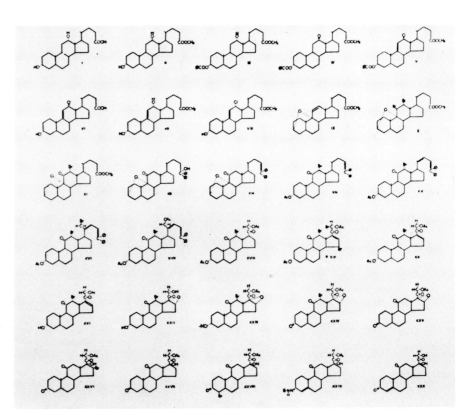

225. Preparation of cortisone. These are the thirty steps used in the Mayo Clinic laboratory to obtain cortisone from a bile acid. Note the step-by-step addition or subtraction of groups of atoms around the central steroid nucleus.

226. Warburg and Fischer. Otto Warburg was perhaps the most accomplished biochemist of all time. His teacher, the great German chemist Emil Fischer, made major contributions to the study of sugars, proteins, and purines. Warburg was a prolific developer of new methodological tools, did extensive research in such areas as photosynthesis and cellular respiration, and found several new enzymes. He considered his most original achievement the discovery in 1924 of the respiratory enzyme iron oxygenase, later known as cytochrome oxygenase. Found in all animal cells, this enzyme plays a crucial role in the transport of oxygen into the cell. The picture was taken at mid-century, on the day the statue of Fischer was unveiled.

227. Warburg's workroom. Surrounded by an array of flasks and test tubes in this modest laboratory setting, Warburg developed methods for investigating the transformation of specific molecules within cells. The century saw the catalytic role of discrete enzymes in metabolic processes claim increasing attention from biochemists, especially in the second quarter. Warburg's techniques for studying tissue respiration underlay much of this work.

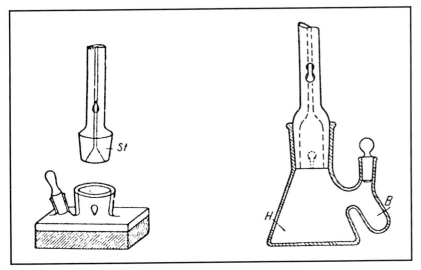

228. Vessels for use with manometric apparatus. Among Warburg's contributions to biochemical research were manometric methods for studying cell metabolism. These illustrations are from a 1930 paper by the German biochemist Hans Adolf Krebs on apparatus for measuring enzymatic breakdown of protein. The container was connected to a manometer (of the type used by Warburg) which monitored, in terms of the change in gas pressure, the amount of gas produced. Krebs spent four years in Warburg's laboratory at Berlin-Dahlem. He later worked out an important metabolic pathway in cells—the sequence of reactions known as the Krebs cycle.

229. Measuring metabolism. "Labeling" a biological compound with a radioactive isotope makes it possible to accurately trace the compound's fate within an animal, plant, or microorganism. By mid-century radioactive tracers had come to play a crucial role in metabolism research. In the experiment pictured here, the urine and feces of a mouse are being collected for analysis. The mouse has eaten foods labeled with such radioisotopes as carbon 14 and hydrogen 3. The amounts of these isotopes in the animal's waste products suggest the rate at which the foods are metabolized. If blood or tissues are directly sampled, the specific fate of atoms of individual foodstuffs can be traced.

230. Isolating chloroplasts. Radioactive tracers helped solve the riddle of photosynthesis. These mid-century photographs show preparations for studying chloroplasts—the tiny cell bodies where photosynthesis occurs. At upper left, spinach leaves are readied for grinding with sand and water. After hand-grinding with a mortar and pestle, the resulting slurry is poured (*above*) through a filter into a flask, which is cooled by ice to preserve the enzymes present. The filtrate, in test tubes, is spun in a centrifuge to separate the chloroplasts from other cell parts. Once isolated, the chloroplasts are transferred, along with radioisotopes, to the reaction vessels with long vents seen at left.

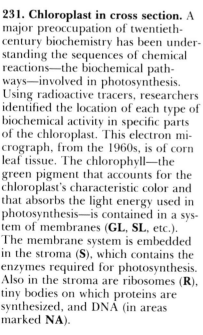

231. Chloroplast in cross section. A major preoccupation of twentieth-century biochemistry has been understanding the sequences of chemical reactions—the biochemical pathways—involved in photosynthesis. Using radioactive tracers, researchers identified the location of each type of biochemical activity in specific parts of the chloroplast. This electron micrograph, from the 1960s, is of corn leaf tissue. The chlorophyll—the green pigment that accounts for the chloroplast's characteristic color and that absorbs the light energy used in photosynthesis—is contained in a system of membranes (**GL**, **SL**, etc.). The membrane system is embedded in the stroma (**S**), which contains the enzymes required for photosynthesis. Also in the stroma are ribosomes (**R**), tiny bodies on which proteins are synthesized, and DNA (in areas marked **NA**).

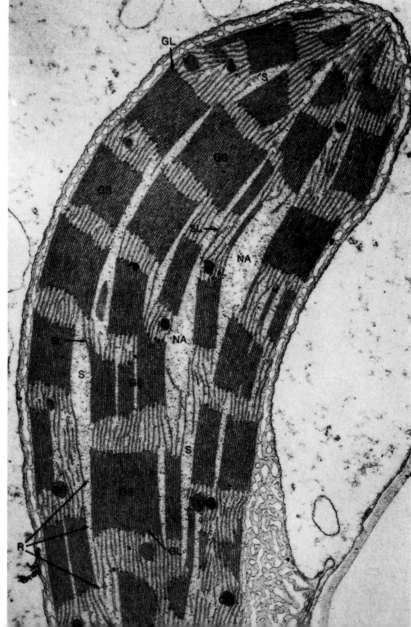

232. Synthesizing protein in the laboratory. The latter decades of the century saw an explosion of interest in the basic genetic material DNA. Biochemists explored the process whereby cells synthesize protein from amino acids in accordance with the genetic information that DNA carries. Experiments like the one shown here were performed in the 1960s. At left, the experimenter grinds colon bacillus cells in a mortar. The "sap" from the ruptured bacteria is still capable of making protein. Below left, this "cell-free system" has been transferred to test tubes; to induce protein synthesis, the experimenter adds stimulants, including synthetic ribonucleic acid (RNA) and amino acids. Below, synthesis is stopped after a specified time interval as trichloroacetic acid is added to the vessel to precipitate out the protein produced.

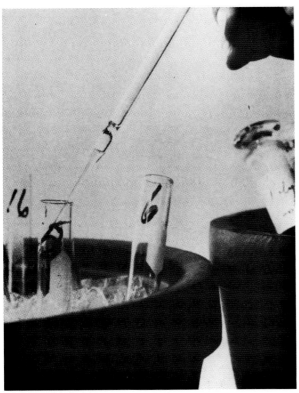

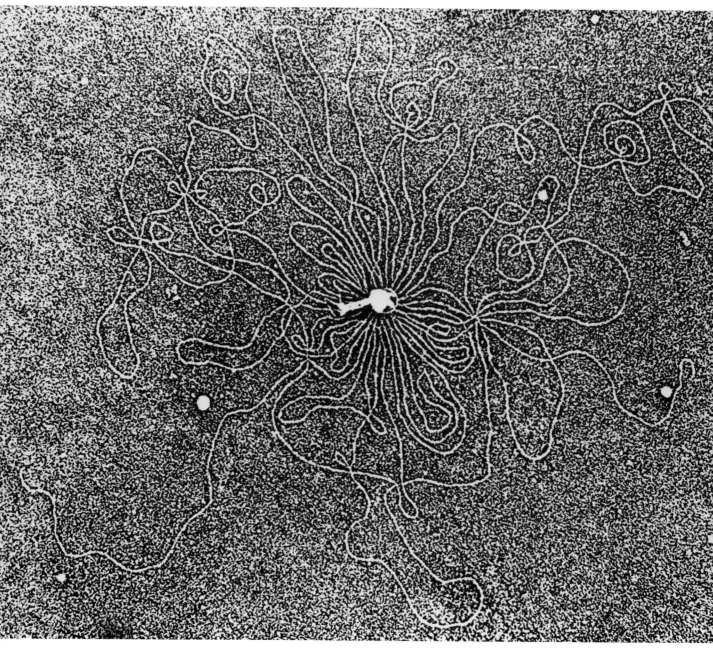

233. DNA spilled out. By a special "osmotic" technique, the head of a virus—a T2 bacteriophage—was ruptured, resulting in this striking 1960 electron micrograph. The DNA, formerly contained in the head, is strewn about the field of view. What we see is actually a single double-stranded thread of DNA; its ends are at bottom left and center right. If the DNA seems too large to fit into the phage, that is because the width of the thread has been expanded tenfold by the procedures used. The "ghost" of the bacteriophage visible in the center of the picture is the virus's protein coat.

16

Molecular Biology

Molecular biology is a new field, the spectacular product of a cluster of breakthroughs around mid-century. It emerged through a consolidation of work in physical chemistry, microbiology, genetics, and biochemistry. Researchers reevaluated the intersecting problems of these traditional fields, as analysis of the structure of the major biological molecules was firmly joined to study of the molecules' roles in the living cell. One by one each of the major biological molecules saw its structure analyzed and its attributes assessed.

The molecule that proved the explanatory power of this approach was the nucleic acid DNA, the material of heredity. For a long time scientists had thought that genes must be protein because only protein molecules seemed to be complex enough to carry complicated information. But the code of heredity turned out to lie in the sequence of bases in a nucleic acid. This startling discovery prompted a reinterpretation of organ, tissue, and cellular research.

The now famous double helix structure of DNA was deciphered in 1953 by a pair of scientists working at Cambridge University's Cavendish Laboratory, which had an extraordinarily rich tradition of Nobel-winning work in physics. They were the British biophysicist Francis Crick and the young American biologist James Watson. Watson and Crick's revolutionary achievement drew on data from X-ray crystallography, a powerful method for exploring molecular structure pioneered early in the century by two English

physicists, William Henry Bragg and his son William Lawrence Bragg. Scientists had used the technique to study crystals of rock salt and diamond, as well as biologically important molecules like hemoglobin, myoglobin, and nucleic acid. Watson and Crick's final model relied heavily on X-ray work from laboratories at Kings College, London, most notably X-ray diffraction data produced by the physical chemist Rosalind Franklin and the biophysicist Maurice Wilkins.

Watson and Crick's recognition that nucleic acid was the material of heredity rested on research in several fields. Important clues to their model came from earlier studies of the genetics of the viruses called phages that infect bacteria, the genetics of the mold *Neurospora*, the inheritance of virulence in pathogenic bacteria, and the chemical composition of DNA.

The study of the inheritance of virulence in bacteria was reported in 1944. Oswald Avery, Colin MacLeod, and Maclyn McCarty at the Rockefeller Institute in New York found that the "transforming principle" (as they called it) was DNA or possibly some other substance adsorbed to DNA. Later, two American biologists, Alfred Hershey and Martha Chase, used radioactive tracers to study the process of phage infection. Their so-called Waring Blendor experiment (phage-infected bacterial cells were spun in a blender) was reported in 1952. It showed that DNA is involved in the replication of phages. Earlier, research on nutrition in bread mold by the

Americans George Beadle and E. L. Tatum in the 1940s had suggested that each gene controlled the production of a single enzyme. Finally, chemical analyses in the late 1940s by the Austrian-born chemist Erwin Chargaff, working at the Columbia College of Physicians and Surgeons, showed that the bases in DNA were present in one-to-one ratios.

Watson and Crick sought a molecular structure for DNA that could encompass and explain these diverse findings. Building on the earlier studies and on structural analysis of the DNA molecule through X-ray crystallography, they proposed a geometric structure that agreed with the empirical data and at the same time possessed the physiological capabilities required of the hereditary material. Their model drew heavily on the "alpha helix" model that the American chemist Linus Pauling had earlier proposed for proteins.

Watson and Crick's molecular model was enormously fruitful in predicting the many kinds of chemical events that could occur in the cell. Watson and Crick said that DNA, which itself could replicate, was transcribed into RNA, which was ultimately translated into protein in the cell. By the late 1950s their analysis had become popularized in the so-called Central Dogma. The one gene–one enzyme relationship proposed by Beadle and Tatum was verified as researchers worked out the DNA codons, or base sequences, that coded for each amino acid in a given protein. In France, François Jacob and Jacques Monod then developed the "operon" theory to explain how genes are turned on and off during the life of the cell.

The molecular models and predictive schemata that followed in the wake of the double helix were given credibility by stunning electron micrographs showing transcription in process. Equally vivid electron micrographs of the multiplication of phages inside the bacterial cell translated chemical events into the visual data so integral to biology. Such images showed minute living organisms invading a cell, reproducing and multiplying, and then bursting the host cell. Radioactive labeling verified the sequence of these events and the relationship between them.

The Watson-Crick discovery opened an era of unprecedented growth for biology. DNA quickly became a household word, as scientists grasped for the first time the molecular mechanism of heredity. In the following decades, researchers—working with new technologies derived from this knowledge and the laboratory techniques associated with it—learned how to modify the mechanism of heredity. Genetic engineering was born. Pieces of biologically important molecules could now be spliced together in new ways. Organisms could now be modified. Ultimately, new cell culture and cell fusion techniques blurred the traditional distinction between plant and animal. New reproductive technologies created life where it had not existed before. Both the test-tube baby and the artificial microorganism were laboratory products of scientists' search for the secret of life in the mechanism of reproduction, which had been laid bare by the new molecular biology.

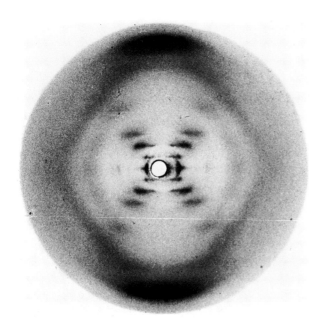

234. X-ray diffraction picture of B form of DNA. The double helix model for the DNA molecule, proposed by James Watson and Francis Crick in 1953, was a momentous achievement for biology. This striking photograph, taken by Rosalind Franklin in May 1952, provided an important stimulus to Watson's thinking when he saw it in early 1953. The Watson-Crick model was in accord with the experimental data obtained by Franklin and her graduate student Raymond Gosling, and they included the photograph in a short paper that was published with Watson and Crick's initial paper. The B form is assumed by the sodium salt of DNA at high humidity.

235. Wilkins. The British biophysicist Maurice Wilkins also sought to decipher DNA's structure. X-ray diffraction studies of DNA that were produced by his laboratory provided crucial data for Watson and Crick, and all three shared the 1962 Nobel Prize in physiology or medicine. The X-ray equipment in the photograph was the type used by Franklin, Gosling, and Wilkins at King's College, London.

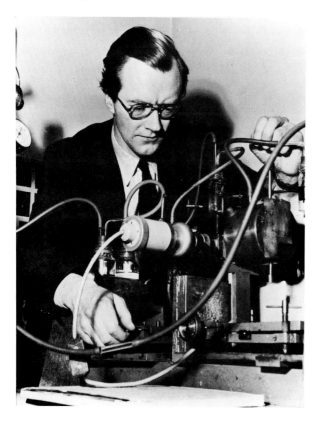

236. Franklin. A colleague of Wilkins's, Rosalind Franklin was a physical chemist and a specialist in X-ray diffraction work. She reportedly was close to solving the structure of DNA when Watson and Crick succeeded in doing it. Franklin died in 1958, four years before the Nobel Prize was awarded for the discovery of the structure of DNA.

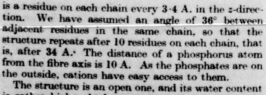

equipment, and to Dr. G. E. R. Deacon and the captain and officers of R.R.S. *Discovery II* for their part in making the observations.

[1] Young, F. B., Gerrard, H., and Jevons, W., *Phil. Mag.*, **40**, 149 (1920).

[2] Longuet-Higgins, M. S., *Mon. Not. Roy. Astro. Soc., Geophys. Supp.*, **5**, 285 (1949).

[3] Von Arx, W. S., Woods Hole Papers in Phys. Oceanog. Meteor., **11** (3) (1950).

[4] Ekman, V. W., *Arkiv. Mat. Astron. Fysik.* (*Stockholm*), **2** (11) (1905).

MOLECULAR STRUCTURE OF NUCLEIC ACIDS

A Structure for Deoxyribose Nucleic Acid

WE wish to suggest a structure for the salt of deoxyribose nucleic acid (D.N.A.). This structure has novel features which are of considerable biological interest.

A structure for nucleic acid has already been proposed by Pauling and Corey[1]. They kindly made their manuscript available to us in advance of publication. Their model consists of three intertwined chains, with the phosphates near the fibre axis, and the bases on the outside. In our opinion, this structure is unsatisfactory for two reasons: (1) We believe that the material which gives the X-ray diagrams is the salt, not the free acid. Without the acidic hydrogen atoms it is not clear what forces would hold the structure together, especially as the negatively charged phosphates near the axis will repel each other. (2) Some of the van der Waals distances appear to be too small.

Another three-chain structure has also been suggested by Fraser (in the press). In his model the phosphates are on the outside and the bases on the inside, linked together by hydrogen bonds. This structure as described is rather ill-defined, and for this reason we shall not comment on it.

We wish to put forward a radically different structure for the salt of deoxyribose nucleic acid. This structure has two helical chains each coiled round the same axis (see diagram). We have made the usual chemical assumptions, namely, that each chain consists of phosphate diester groups joining β-D-deoxyribofuranose residues with 3′,5′ linkages. The two chains (but not their bases) are related by a dyad perpendicular to the fibre axis. Both chains follow right-handed helices, but owing to the dyad the sequences of the atoms in the two chains run in opposite directions. Each chain loosely resembles Furberg's[2] model No. 1; that is, the bases are on the inside of the helix and the phosphates on the outside. The configuration of the sugar and the atoms near it is close to Furberg's 'standard configuration', the sugar being roughly perpendicular to the attached base. There

This figure is purely diagrammatic. The two ribbons symbolize the two phosphate—sugar chains, and the horizontal rods the pairs of bases holding the chains together. The vertical line marks the fibre axis

is a residue on each chain every 3·4 A. in the z-direction. We have assumed an angle of 36° between adjacent residues in the same chain, so that the structure repeats after 10 residues on each chain, that is, after 34 A. The distance of a phosphorus atom from the fibre axis is 10 A. As the phosphates are on the outside, cations have easy access to them.

The structure is an open one, and its water content is rather high. At lower water contents we would expect the bases to tilt so that the structure could become more compact.

The novel feature of the structure is the manner in which the two chains are held together by the purine and pyrimidine bases. The planes of the bases are perpendicular to the fibre axis. They are joined together in pairs, a single base from one chain being hydrogen-bonded to a single base from the other chain, so that the two lie side by side with identical z-co-ordinates. One of the pair must be a purine and the other a pyrimidine for bonding to occur. The hydrogen bonds are made as follows: purine position 1 to pyrimidine position 1; purine position 6 to pyrimidine position 6.

If it is assumed that the bases only occur in the structure in the most plausible tautomeric forms (that is, with the keto rather than the enol configurations) it is found that only specific pairs of bases can bond together. These pairs are: adenine (purine) with thymine (pyrimidine), and guanine (purine) with cytosine (pyrimidine).

In other words, if an adenine forms one member of a pair, on either chain, then on these assumptions the other member must be thymine; similarly for guanine and cytosine. The sequence of bases on a single chain does not appear to be restricted in any way. However, if only specific pairs of bases can be formed, it follows that if the sequence of bases on one chain is given, then the sequence on the other chain is automatically determined.

It has been found experimentally[3,4] that the ratio of the amounts of adenine to thymine, and the ratio of guanine to cytosine, are always very close to unity for deoxyribose nucleic acid.

It is probably impossible to build this structure with a ribose sugar in place of the deoxyribose, as the extra oxygen atom would make too close a van der Waals contact.

The previously published X-ray data[5,6] on deoxyribose nucleic acid are insufficient for a rigorous test of our structure. So far as we can tell, it is roughly compatible with the experimental data, but it must be regarded as unproved until it has been checked against more exact results. Some of these are given in the following communications. We were not aware of the details of the results presented there when we devised our structure, which rests mainly though not entirely on published experimental data and stereochemical arguments.

It has not escaped our notice that the specific pairing we have postulated immediately suggests a possible copying mechanism for the genetic material.

Full details of the structure, including the conditions assumed in building it, together with a set of co-ordinates for the atoms, will be published elsewhere.

We are much indebted to Dr. Jerry Donohue for constant advice and criticism, especially on interatomic distances. We have also been stimulated by a knowledge of the general nature of the unpublished experimental results and ideas of Dr. M. H. F. Wilkins, Dr. R. E. Franklin and their co-workers at

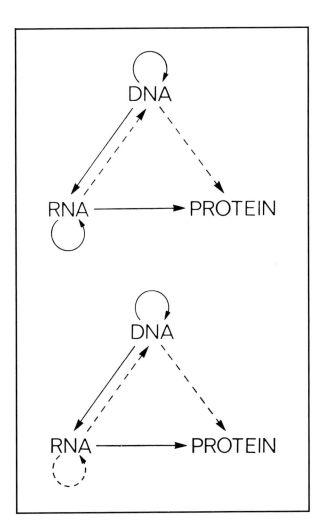

238. Central dogma of molecular biology. The information, or genetic code, carried by the nucleic acid DNA prescribes how amino acids are to be joined together to build proteins. An intermediary in the process is another nucleic acid, RNA. (In some viruses, however, the fundamental nucleic acid is RNA.) In 1958, Crick summed up this situation in a principle he called the Central Dogma: "once 'information' has passed into protein *it cannot get out again*. In more detail, the transfer of information from nucleic acid to nucleic acid, or from nucleic acid to protein may be possible, but transfer from protein to protein, or from protein to nucleic acid is impossible." The top triangle, from a paper published by Crick in 1970, represents his 1958 views of the transfer relations between the three key molecules—DNA, RNA, and protein. Solid arrows indicate probable transfers; dashed arrows, those thought to be possible. The bottom triangle shows Crick's revised views in 1970. Here, solid arrows show "general" transfers, occurring in all cells; dashed arrows are rare "special" transfers, occurring only in special circumstances.

237. Watson-Crick paper. This page from the British journal *Nature* contains virtually all of Watson and Crick's first published statement of their double helix structure for DNA. Near the end the authors dryly remark that the structure's implications for a replication mechanism for the genetic material "has not escaped our notice." This observation about replication possibilities has been called one of the coyest statements in the literature of science.

239. Replication of DNA. How does DNA replicate itself? The answer given by the double helix model is that the two strands unwind, and each serves as a template for the construction of a new partner strand. Each strand features a long chain of four different kinds of nitrogen bases, symbolized A, G, C, and T. The two strands are held together by relatively weak bonds between complementary bases. Opposite an A in one strand there can be only a T in the other, and opposite a G only a C. Thus, a new strand formed opposite the untwisted template strand will be a copy of the template's original partner.

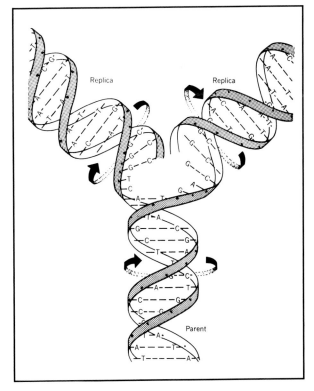

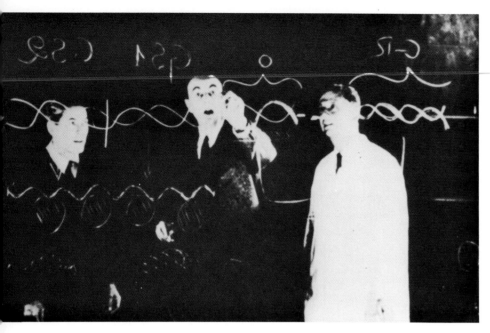

240. Jacob, Monod, and Lwoff. This photograph of the French scientists François Jacob (*left*), Jacques Monod (*center*), and André Lwoff was taken at the Pasteur Institute after the announcement of their 1965 Nobel Prize. Jacob and Monod suggested that in bacteria a form of RNA called messenger RNA carries the information encoded in DNA to the sites of protein synthesis and that genes directing the synthesis of this messenger RNA exist in gene complexes called operons that include a controlling operator gene. Lwoff's work focused on the mechanism of latent infection (lysogeny) of bacteria by viruses.

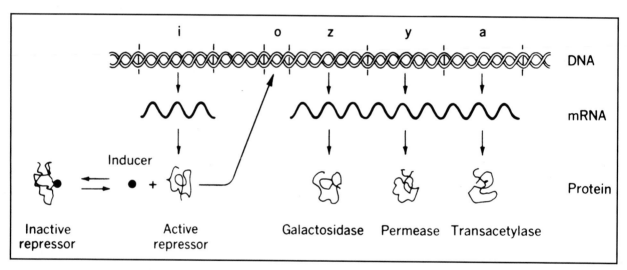

241. Operon theory. Here Jacob and Monod's operon concept is applied to the genes of the colon bacillus *Escherichia coli*. Genes **z**, **y**, and **a** code for enzymes involved in the fermentation of lactose, or milk sugar (the enzymes galactosidase, permease, and transacetylase). When lactose, the "inducer" substance, is present, the enzymes are needed, and the genes coding for them swing into action. These genes are controlled by the operator gene **o**. A separate gene, **i**, codes for a repressor substance, which can attach to **o**, closing it. If lactose is present, it can combine with the repressor and inactivate it, thus keeping it from switching off **o**.

242. Genes at work. This picture, ▶ made in 1969, is probably the first electron micrograph of DNA in the act of spinning out fibrils of RNA. The DNA came from a salamander reproductive cell. The RNA being produced is for the ribosomes, the sites of protein synthesis. The DNA is thus transcribing ribosomal RNA.

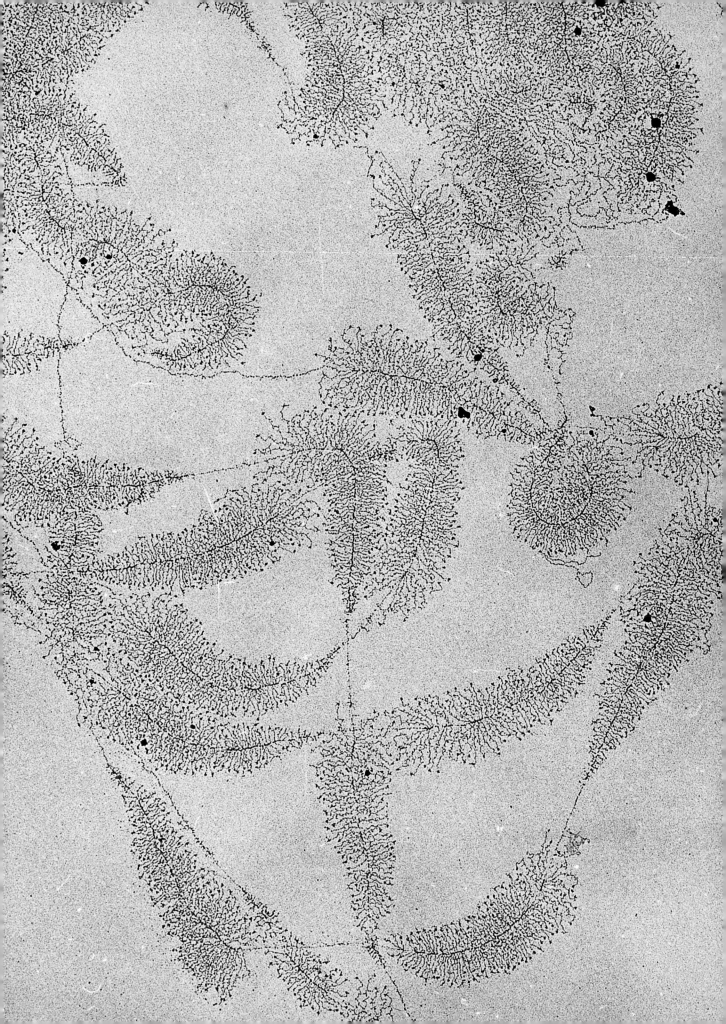

243. Prehistory of molecular genetics: Beadle and his mold. Focusing on a relatively simple organism, the American geneticist George Beadle, together with the biochemist Edward Tatum, reported in 1941 the discovery that in the red bread mold, *Neurospora crassa*, a cell's production of an enzyme was controlled by a single gene: one gene, one enzyme was the basic principle. In the 1950s the double helix model of DNA would firmly link biochemistry and genetics. In the photograph Beadle (*center*) is studying some of his thousands of test tubes containing strains of the mold; the project required a large team of researchers.

244. DNA unmasked. In a landmark 1944 paper the Canadian-American bacteriologist Oswald Avery and his colleagues Colin MacLeod and Maclyn McCarty reported evidence that DNA is the carrier of heredity. Below are two strains of pneumonia-causing bacteria. At left is a small, rough-surfaced, and harmless pneumococcus that lacks a protective capsule. At right are larger, smooth-surfaced bacteria having a protective capsule and the power to infect. The "rough" strain can be made to regain its capsules and virulence through inoculation with heat-killed bacteria of a smooth strain. This transformation is then inherited. The transforming agent in the heat-killed cells, suggested Avery and his coworkers, is DNA.

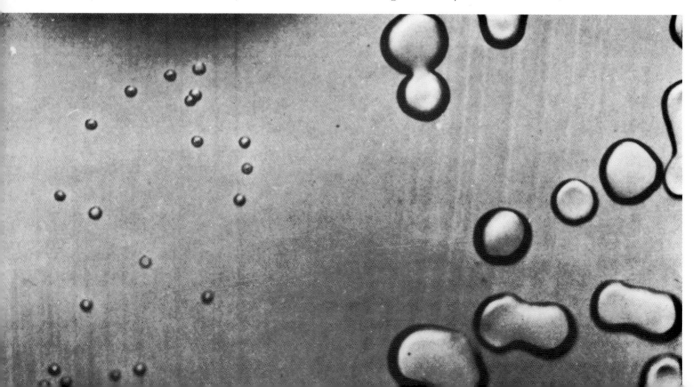

245. Bacteriophages: an early view.
When it was trained on the bacteria-infecting viruses called bacteriophages, the electron microscope showed them to be tadpole-shaped bodies. This 1942 micrograph includes so-called T-even bacteriophages together with *Escherichia coli* bacteria. Subsequent research on bacteriophages provided strong supporting evidence for Avery's suggestion that DNA is the basic genetic material: when bacteriophages infect a bacterium, it was found, DNA from the bacteriophage is injected into the bacterium, where the genetic information in that DNA directs the production of new bacteriophages.

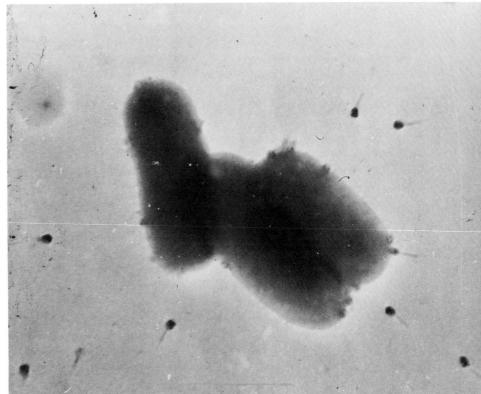

246. Triggered bacteriophages. This striking later micrograph of T-even phages catches them in their triggered state. Their hexagonal heads are empty, the DNA inside having been injected into the bacterial host. Clearly visible are the bacteriophages' contracted tail sheaths and hollow tail needles.

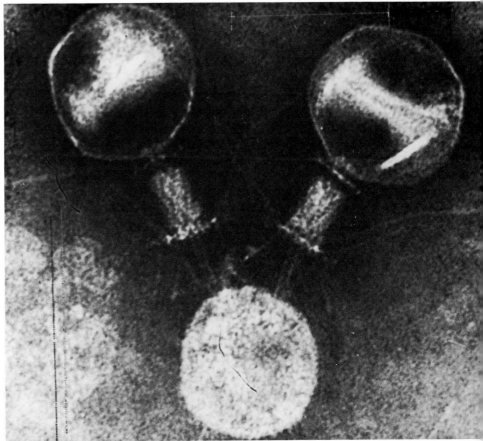

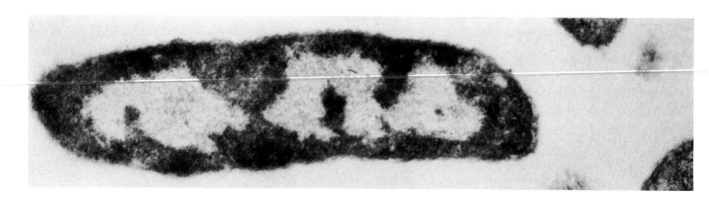

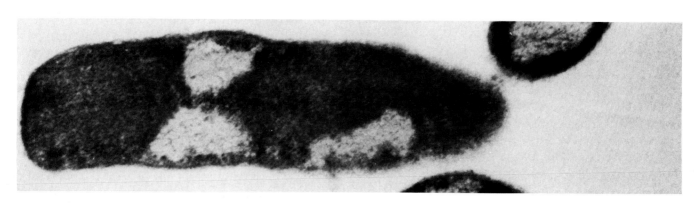

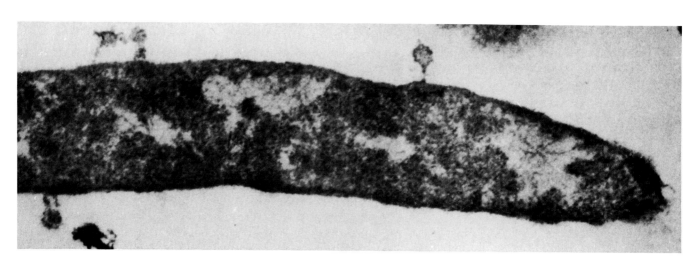

247. Bacteriophages invade a bacterium. This classic series of electron micrographs shows what happens when the bacterium *Escherichia coli* is infected with T2 phages. At top we see the bacillus before infection. The second picture was taken four minutes after infection; cavities, or vacuoles, are forming along the cell wall. In the third picture, ten minutes after infection, the interior of the cell has been reorganized; there are pools of new viral components. The series continues on the next page.

248. The phages multiply. Here are more images from the series depicting the growth of viruses inside the bacterial host. The first micrograph, taken twelve minutes after infection, shows new phages starting to form. In the second picture, half an hour has elapsed since infection; the bacterial cell contains more than fifty fully developed T2 viruses and will soon burst open, releasing them. These remarkable micrographs were made by Edouard K. Kellenberger of the University of Geneva.

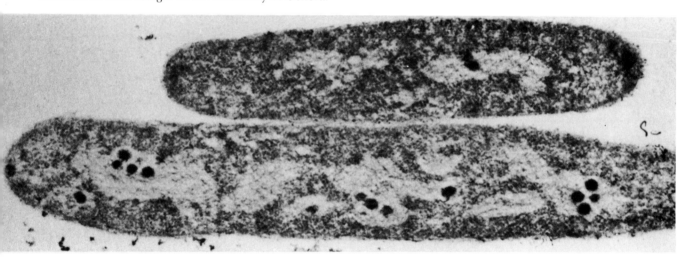

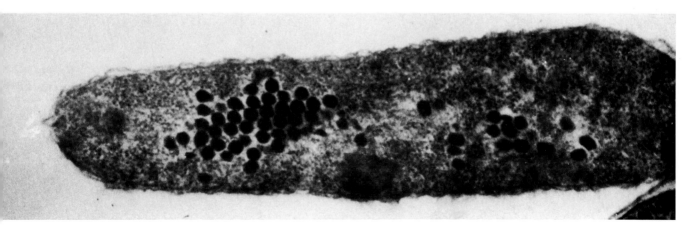

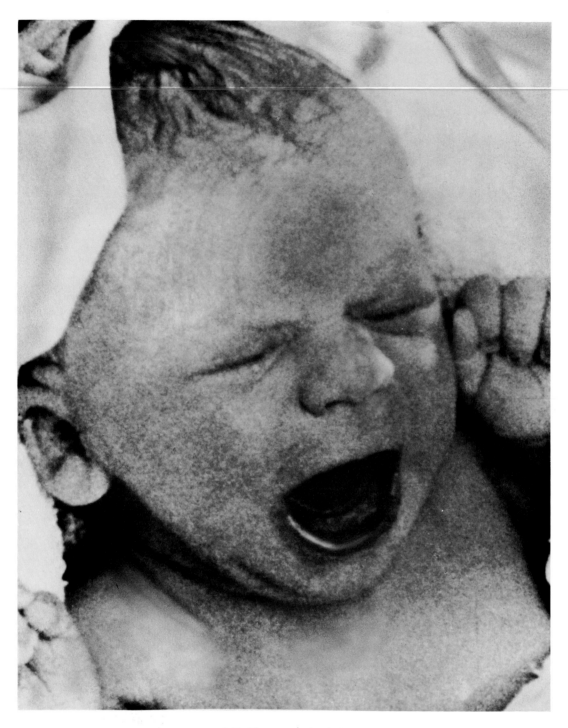

249. First test-tube baby. Louise Jay Brown was born on July 25, 1978, in a hospital in Oldham, England. Her conception took place in "in vitro," in a laboratory dish, where eggs plucked from her mother's ovaries were fertilized by sperm from her father. (The mother could not conceive, because of blocked fallopian tubes.) One of the fertilized eggs was inserted in the mother's womb. Other techniques of human reproductive technology in the latter decades of the century included the freezing and storing of reproductive cells and the carrying of a fetus to term by a "surrogate" mother.

17

Origin and Procreation of Life

In the latter decades of the nineteenth century chemical analysis of the constituents of living organisms produced increasing speculation about the origin and nature of protoplasm, the chemical substance of the living cell. At the same time, such major figures in the rise of evolutionary theory as Charles Darwin and Thomas Henry Huxley commented on the possible origin of life in the primordial seas. Darwin believed an environment much different from today's was required. Consider his words, in an 1871 letter to a friend: "But if (and oh what a big if) we could conceive in some warm little pond, with all sorts of ammonia and phosphoric salts, light, heat, electricity, etc. present, that a protein compound was chemically formed, ready to undergo still more complex changes, at the present day such matter would be instantly devoured, or absorbed, which would not have been the case before living creatures were formed."

At the turn of the twentieth century many scientists still thought that protoplasm was a single substance formed by a highly complex molecule. In the early decades, however, it was shown to consist of many diverse compounds contained in different regions of the cell. As chemical knowledge of the cell rapidly increased, so did researchers' interest in the origins of life. Later scientists elaborated on their predecessors' ideas, setting forth more fully developed theories of how life might have begun and explicitly devising experimental techniques to test these hypotheses. Such

work became especially popular at mid-century as the evolutionary and biochemical approaches converged, bringing all the life sciences under a common theoretical framework.

In 1953 the American biochemist Stanley Miller, then a graduate student, created amino acids—the building blocks of proteins—by discharging an electric spark through a mixture of water, methane, hydrogen, and ammonia, a mixture thought to approximate the compounds available in the ancient atmosphere. This suggested how life might have originated on the ancient earth. Miller's observations stimulated further discussion of possible synthetic processes in the primeval environment. Other researchers studied the structure and behavioral properties of chemical systems—such as coacervates and proteinoid microspheres—that resembled primitive forms of life. Underlying much of this work were essays by the Soviet biochemist Aleksandr Ivanovich Oparin on the gradual evolution of organic substances (published under the title *Proiskhozhdenie zhizni* [Origin of Life] in 1924; a fuller presentation of his thought appeared in 1936, with an English translation in 1938).

Later in the century, analyses of DNA and RNA in a variety of present-day species suggested to some that the earliest life-form was a single-celled sulfur-consuming organism dwelling in near-boiling water (as in the hydrothermal vents of the ocean bottom). Others argued that the origins of life were to be sought in the heavens. Organic molecules were detected beyond the earth—in interstellar dust, in comets. Scientists found amino acids in meteorites, most notably in the Murchison meteorite, which landed in Australia in 1969.

Organisms' ability to grow and reproduce—the ability of like to beget like—remained a major concern of experimental embryology and genetics in the second half of the century. But the discussion came to be oriented in a new way. The rise of molecular genetics and molecular biology focused attention on the properties of nucleic acids and enzyme systems. At the same time, scientists' newly acquired capability to maintain cells in artificial media made possible analyses of growth and differentiation in cultures of isolated cells. The result was increased biological understanding of how specific form is maintained from one generation to another, as well as increased capability to control developmental processes.

These advances were not just of academic interest: they had major commercial consequences as well. With the new cell culture techniques, specific types of cells could be separated from related cells and cultivated in controlled environments. These cultured cells could be used to produce chemical substances, new "biologicals," of therapeutic value to medicine. Culture techniques also became important in agriculture. Here the ability to reproduce good-quality, uniform organisms was highly prized. Genetically identical plants were produced by cloning—that is, the asexual reproduction of cells. Researchers explored similar techniques for stock animals.

Eventually, through these manipulative cell culture techniques, biologists constructed new kinds of organisms. For example, in the 1950s they began to transplant cell nuclei in order to study the interplay between nuclear and cytoplasmic factors in cell development. Researchers subsequently grew animals from eggs with transplanted nuclei. Plant-animal hybrids were also produced.

Beginning in the 1970s, molecular biology introduced yet another method for experimental study of "mixed" organisms. New "gene-splicing" techniques enabled scientists to join together genetic information from different organisms. This technology created the possibility of mass-producing genetically modified organisms. Vigorous debate ensued among scientists and the public at large, as people raised questions about the potential dangers posed by such organisms. Despite the controversy, development of manipulative laboratory procedures continued, and commercial possibilities grew. The 1980s saw new industrial applications, as well as the production of pharmaceutically important molecules like insulin and interferon. New organisms created by gene-splicing techniques were even patented. Unanticipated ethical and legal questions multiplied, however. Could, for example, the openness of science be maintained, and public safety preserved, in an atmosphere of commercially motivated, secret research?

Perhaps the most controversial of the new procedures introduced in the latter part of the century were the human reproductive technologies, including artificial insemination with stored sperm, "in vitro" fertilization, and surrogate motherhood. These technologies originated from earlier twentieth-century research on problems of infertility in animals and human beings. The apparent applicability of the procedures to human fertility problems made them appealing to many people in industrialized countries. But these high-tech solutions to human problems created new problems. The advent of the new reproductive technologies triggered considerable public discussion of the social impact of modern biological research, as well as the ethics of this enterprise.

The new reproductive technologies also gave rise to new questions, vigorously voiced, about the limits of human understanding and the hubris of the modern scientific endeavor. Laboratory research had begun to challenge long-established assumptions about the place of human beings in the natural order. Human intervention in the laboratory had provoked a series of perturbations of that order. As a result, public perceptions of the potential fruits of laboratory research shifted noticeably, from unbridled enthusiasm to wariness of unforeseen side effects having perhaps perilous consequences.

Industrialized biological research lay at the heart of the debate. The technological "brave new world" predicted by the English novelist Aldous Huxley in 1932 was coming into being. The main features of the increasingly industrialized environment of biological research will be explored in Chapters 26 and 27.

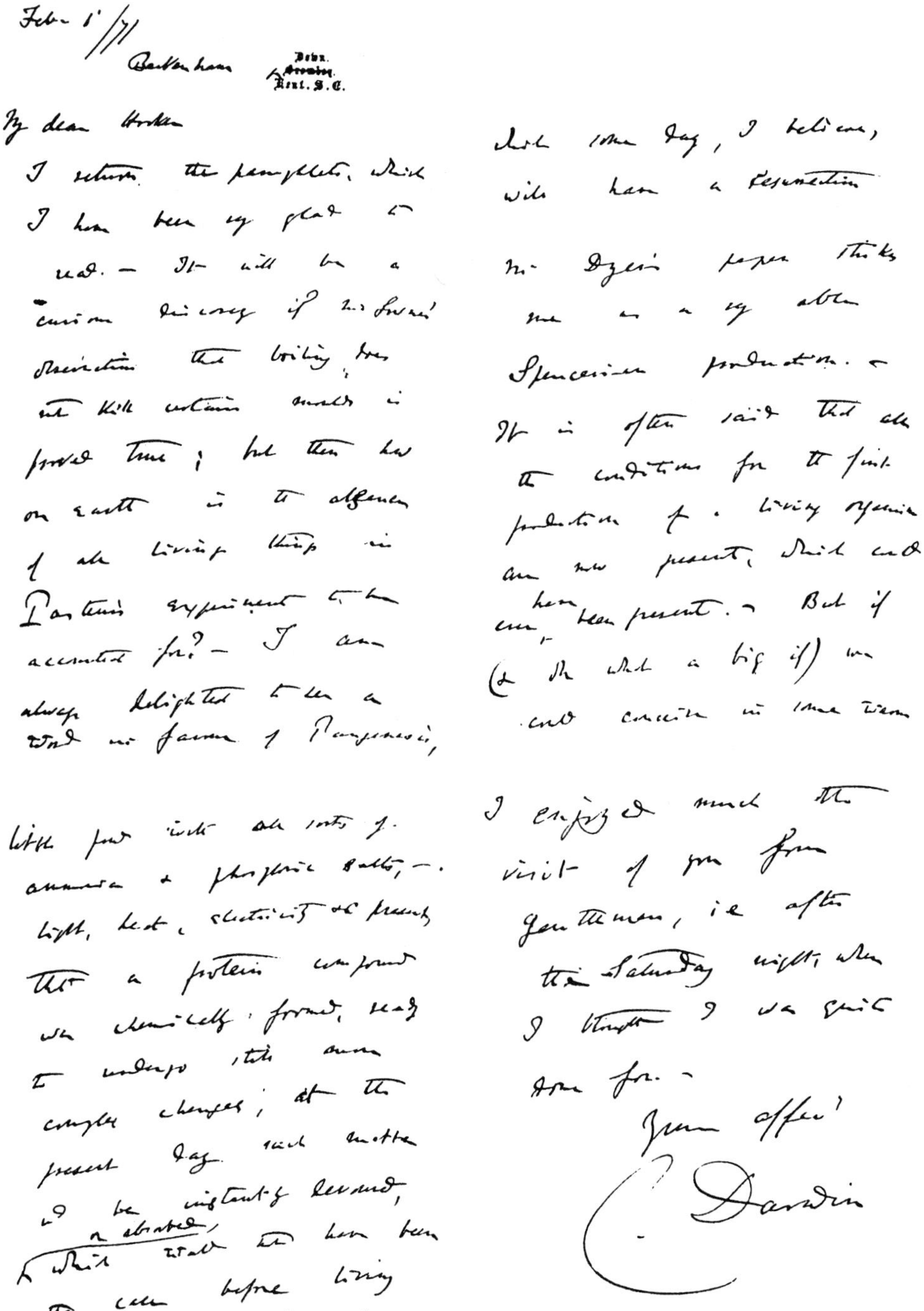
Feb 1/71

Beckenham
Down.
~~Bromley.~~
Kent. S.E.

My dear Hooker

I return the pamphlet, which I have been very glad to read.— It will be a curious discovery if Dr Sanderson's observation the boiling does not kill certain moulds is proved true; but then how on earth is the absence of all living things in Pasteur's experiment to be accounted for?— I am always delighted to see a word in favour of Pangenesis, which some day, I believe, will have a resurrection.

Mr Dyer's paper strikes me as a very able Spencerian production.—

It is often said that all the conditions for the first production of a living organism are now present, which could ever have been present.— But if (& oh what a big if) we could conceive in some warm little pond with all sorts of ammonia & phosphoric salts,—light, heat, electricity &c present, that a protein compound was chemically formed, ready to undergo still more complex changes, at the present day such matter wd be instantly devoured, or absorbed, which would not have been the case before living creatures were formed.—

Henrietta makes hardly any progress, & God knows when she will be well.—

I enjoyed much the visit of you four gentlemen, i.e. after the Saturday night, when I thought I was quite done for.—

Yours affect

C. Darwin

250. Darwin on the origins of life. Charles Darwin suggested that life could have begun spontaneously from available chemicals in the distant past in an environment radically different from today's. A "warm little pond" might have served, he suggests in this 1871 letter to a friend. But Darwin saw that a sterile environment was required for the first organic compounds to persist, because today "such matter would be instantly devoured, or absorbed."

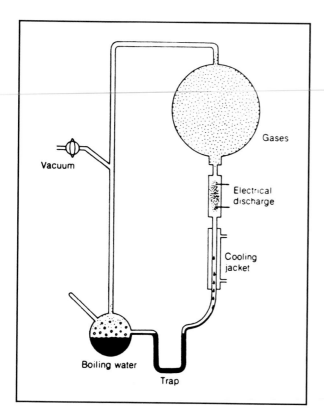

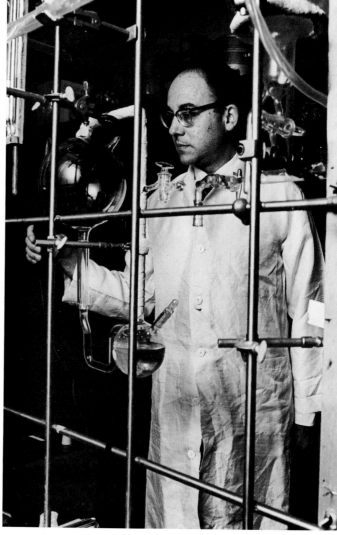

251. Miller's experiment. In 1953 the American biochemist Stanley Miller demonstrated that complex molecules of the kind now found in living organisms could have been produced spontaneously in the conditions presumed to exist on the primitive earth. Miller circulated a mixture of simple gases past an electric spark discharge, representing flashes of lightning. After several days of sparking, he found amino acids in his solution. Paper chromatography was used to detect the amino acids, collected from the trap at the bottom of the apparatus.

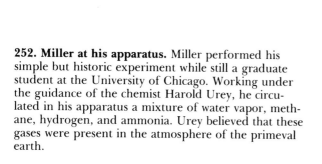

252. Miller at his apparatus. Miller performed his simple but historic experiment while still a graduate student at the University of Chicago. Working under the guidance of the chemist Harold Urey, he circulated in his apparatus a mixture of water vapor, methane, hydrogen, and ammonia. Urey believed that these gases were present in the atmosphere of the primeval earth.

253. Oparin's coacervates. In the 1920s the Soviet biochemist A. I. Oparin outlined the view of how life originated that found support in Miller's experiment three decades later. According to Oparin, life gradually developed in the ocean of the early earth, whose oxygenless atmosphere contained such gases as ammonia, hydrogen, and methane—much more conducive than oxygen to the synthesis and preservation of many organic molecules. (About the same time the British biologist J. B. S. Haldane suggested that a primordial "soup" of organic molecules arose in the ancient seas.) As a model of how cells might have developed, Oparin studied the behavior of coacervate droplets, organized clumps of complex colloidal molecules exhibiting some properties associated with living organisms. Coacervates tend to concentrate certain substances inside themselves, rejecting others. The coacervates shown here were formed in an aqueous solution of protamine (a type of protein) and polyadenylic acid (RNA).

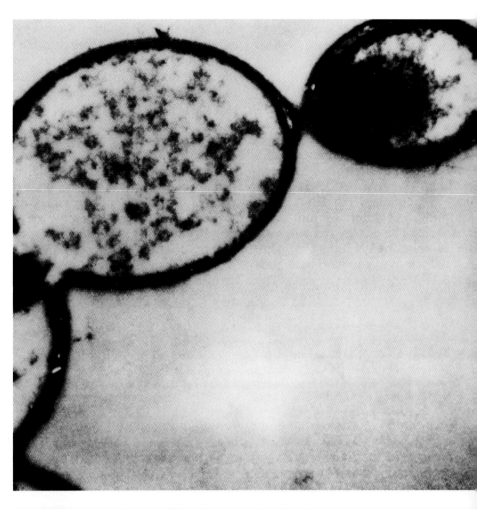

254. Thermal proteinoid microspheres. Coacervates were not the only structures studied by researchers on the origin of life. The American biochemist Sidney Fox focused on thermal proteinoid, a polymer produced by dry mixtures of amino acids when subjected to moderate heat. Under certain conditions this polymer can form slowly growing microspheres, which in time bud.

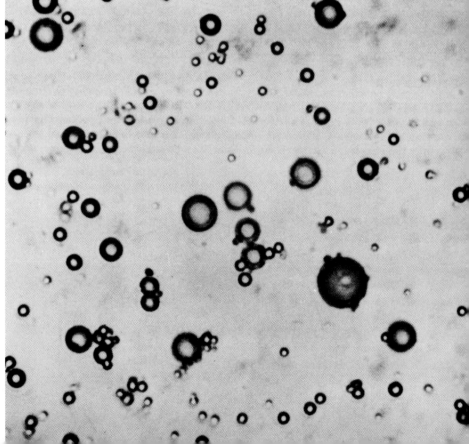

255. Reproductive technology: plant protoplasts. In the 1970s the cloning, or asexual reproduction, of plants from protoplasts became an active area of research with practical implications for agriculture. Protoplasts are individual plant cells stripped of their outer wall. The ones we see here are from potato plant leaf cells. In a culture medium, the protoplasts, which contain a nucleus and chloroplasts, make new cell walls, grow, and divide. Eventually entire plants are regenerated.

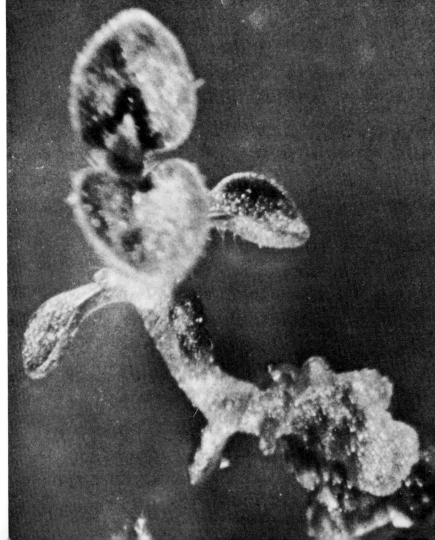

256. A new potato shoot, from cloned cells. After six weeks of incubation in a special medium, shoots rise out of the masses of cells that developed from the leaf-cell protoplasts. Such clones were used to grow fields of genetically identical potato plants and to study the process of genetic mutation.

257. Tadpole, from an egg with transplanted nucleus. Regenerating animals rather than plants from single cells proved a task less amenable to solution. A step forward was made in 1953 by Robert W. Briggs and Thomas J. King of the Institute for Cancer Research in Philadelphia. They succeeded in growing tadpoles from eggs whose nucleus had been replaced with a nucleus from a partly differentiated cell of a developing embryo.

258. Mice, from eggs with transplanted nucleus. By the beginning of the 1980s scientists succeeded in making viable mammals through nuclear transplantation. These mice are the proof. They developed from fertilized white mouse eggs whose nuclei were replaced with nuclei from gray mouse embryo cells.

259. Hybrid plant-animal cell. In the mid-1970s scientists at the Brookhaven National Laboratory constructed this cell containing nuclei from both plant and animal cells. The small dark spots are nuclei from tobacco plant cells; the large spots are nuclei from human HeLa tumor cells.

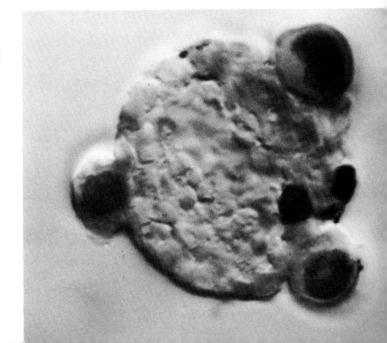

DEPARTMENT OF MOLECULAR BIOPHYSICS
AND BIOCHEMISTRY

July 17, 1973

Dr. Philip Handler, President
National Academy of Sciences
2101 Constitution Avenue
Washington, DC 20418

Dear Doctor Handler:

We are writing to you, on behalf of a number of scientists, to communicate a matter of deep concern. Several of the scientific reports presented at this year's Gordon Research Conference on Nucleic Acids (June 11-15, 1973, New Hampton, New Hampshire) indicated that we presently have the technical ability to join together, covalently, DNA molecules from diverse sources. Scientific developments over the past two years make it both reasonable and convenient to generate overlapping sequence homologies at the termini of different DNA molecules. The sequence homologies can then be used to combine the molecules by Watson-Crick hydrogen bonding. Application of existing methods permits subsequent covalent linkage of such molecules. This technique could be used, for example, to combine DNA from animal viruses with bacterial DNA, or DNAs of different viral origin might be so joined. In this way new kinds of hybrid plasmids or viruses, with biological activity of unpredictable nature, may eventually be created. These experiments offer exciting and interesting potential both for advancing knowledge of fundamental biological processes and for alleviation of human health problems.

Certain such hybrid molecules may prove hazardous to laboratory workers and to the public. Although no hazard has yet been established, prudence suggests that the potential hazard be seriously considered.

A majority of those attending the Conference voted to communicate their concern in this matter to you and to the President of the Institute of Medicine (to whom this letter is also being sent). The conferees suggested that the Academies establish a study committee to consider this problem and to recommend specific actions or guidelines should that seem appropriate. Related problems such as the risks involved in current large-scale preparation of animal viruses might also be considered.

A list of participants in the Conference is attached for your interest.

Sincerely yours,

Maxine Singer and Dieter Söll (us)

Maxine Singer
National Institutes of Health
Room 9N-119, Building 10
Bethesda, MD 20014

Dieter Soll
Associate Professor of Molecular Biophysics
Yale University
New Haven, CT 06520

Maxine Singer
Dieter Soll
Co-Chairmen of the 1973 Gordon
  Research Conference on Nucleic Acids

Enclosure

260. Genetic engineering: the potential for hazard. Scientists discovered in the early 1970s that pieces of DNA from different sources could be readily joined together to form hybrid genetic material. This was an exciting development, potentially heralding the creation of new therapeutic agents. But it also raised the specter of the inadvertent or intentional creation of hazardous organisms. In July 1974 a U.S. National Academy of Sciences committee called on world scientists to halt two types of research in order to assess the risks involved. Three months later the U.S. National Institutes of Health set up a committee to evaluate the potential hazards of DNA recombinants and to develop guidelines for researchers.

261. Research proceeds, with caution. By mid-1976, after two years of a moratorium by some scientists on certain forms of recombinant DNA research, the U.S. National Institutes of Health issued guidelines (mandatory only for institutions receiving federal funds) for the work to move ahead. Among other things, the guidelines sought to limit research to organisms unlikely to survive outside the laboratory, and to make sure that newly created organisms would not escape from laboratory confinement. With the United States playing a leading role in recombinant DNA research, the American guidelines were distributed abroad, in hopes of securing international cooperation to avoid the potential dangers of the research.

Department of State **TELEGRAM**

UNCLASSIFIED 7766

DRAFTED BY HEW/NIH:JRQUINN
APPROVED BY OES/APT/BHP:WJWALSH, III
NEA/EX:EGABINGTON(INFO)
ARA/EX:GECHAFIN(INFO)
FA/EX:JEFFREY CUNNINGHAM(INFO)
EUR/EX:CEREDMAN(INFO)
AF/EX:TFORD(INFO)
DHEW/OIH-JRKING

------------------------ 040862

R 222057Z JUN 76
FM SECSTATE WASHDC
TO AMEMBASSY ANKARA
AMEMBASSY ATHENS   AMEMBASSY CARACAS   AMEMBASSY PRETORIA
AMEMBASSY BELGRADE   AMEMBASSY COPENHAGEN   AMEMBASSY ROME
AMEMBASSY BEIRUT BY POUCH AMEMBASSY DUBLIN   AMCONSUL RIO DE JANEIRO
AMEMBASSY BERLIN   USMISSION GENEVA   AMEMBASSY SAN JOSE
AMEMBASSY BERN   AMEMBASSY THE HAGUE   AMEMBASSY SANTIAGO
AMEMBASSY BOGOTA   AMEMBASSY HELSINKI   AMEMBASSY SOFIA
AMEMBASSY BONN   AMEMBASSY LISBON   AMEMBASSY STOCKHOLM
AMEMBASSY BRASILIA   AMEMBASSY MADRID   AMEMBASSY TAIPEI
AMEMBASSY BRUSSELS   AMEMBASSY MONTEVIDEO   AMEMBASSY TEL AVIV
AMEMBASSY BUDAPEST   AMEMBASSY MOSCOW   AMEMBASSY TOKYO
AMEMBASSY BUENOS AIRES   AMEMBASSY OSLO   AMEMBASSY VIENNA
AMEMBASSY CAIRO   AMEMBASSY OTTAWA   AMEMBASSY WELLINGTON
AMEMBASSY CANBERRA   AMEMBASSY PARIS   USMISSION USUN NEW YORK
                         USLO PEKING   USMISSION OECD PARIS
                         AMEMBASSY PRAGUE   AMCONSUL CAPE TOWN

UNCLAS STATE 154658

SUBJECT: NIH GUIDELINES ON DNA/RNA(GENETIC ENGINEERING ETC.) RECOMBINANT RESEARCH

1. NATIONAL INSTITUTES OF HEALTH PLANS ANNOUNCE GUIDELINES ON JUNE 23, 1976, CONCERNING RECOMBINANT DEOXYRIBO NUCLEIC ACID(DNA) RESEARCH DEVELOPED AS A RESULT OF OPEN DISCUSSIONS WITH U.S. SCIENTISTS.

2. IN VIEW OF SERIOUS IMPORTANCE OF SUBJECT TO WORLD'S SCIENTISTS AND NEED FOR WORLD PUBLICITY AND COOPERATION ON PROBLEM, DEPARTMENT REQUESTS ADDRESSEE MISSIONS ALERT APPROPRIATE OFFICIALS OF HOST GOVERNMENT'S AGENCY MOST CONCERNED WITH SCIENCE AND/OR SUPPORT OF BIOMEDICAL RESEARCH AND OFFER TO PROVIDE COPY OF GUIDELINES SOONEST. COMMENTS ON GUIDELINES MAY BE ADDRESSED TO DIRECTOR, NATIONAL INSTITUTES OF HEALTH, BETHESDA, MARYLAND 20014.

3. TWO COPIES OF THE GUIDELINES WILL BE AIRPOUCHED SEPA- RATELY TO POST SHORTLY AFTER RELEASE DATE OF JUNE 23. APPRECIATE DELIVERY OF GUIDELINES ASAP.

4. NIH ALSO PLANS SEND COPY GUIDELINES DIRECTLY TO ALL GRANTEES AND CONTRACTORS, DOMESTIC AND FOREIGN.

5. PLEASE NOTIFY NIH (ATTN: DR. J.R. QUINN) OF PERSON IN HOST GOVERNMENT ADVISED OF ABOVE MESSAGE AND TO WHOM GUIDELINES DELIVERED. ROBINSON

262. Results of genetic engineering: plant survival. A major goal of genetic engineering research has been to increase the resistance of plants to herbicides. In petunias, for example, the herbicide glyphosate suppresses the activity of a vital enzyme. In the mid-1980s experiment pictured here, genetic material increasing production of the enzyme was introduced into petunias, enabling them to fluorish (*top*) in the presence of amounts of herbicide that were enough to kill ordinary petunias (*bottom*).

Part Five

THE
HUMAN
ANIMAL

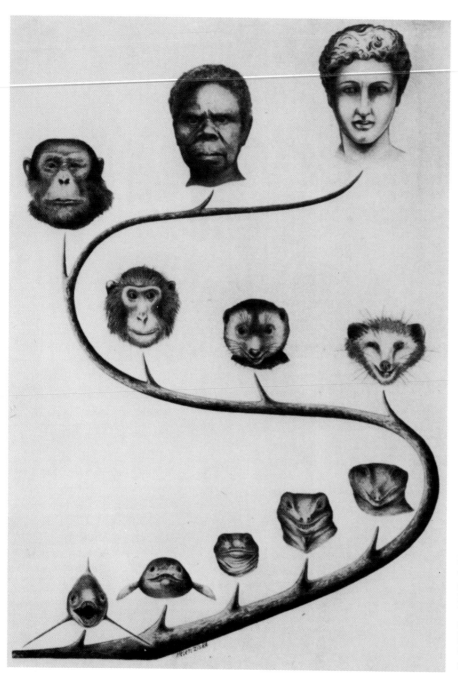

263. Our face from fish to man. This charming summary of a hypothetical evolutionary path is from a 1929 book by the American paleontologist William King Gregory. The path runs from Devonian shark through lobe-finned fish, amphibian, Permo-Carboniferous reptile, Triassic mammal-like reptile, Cretaceous mammal, lemuroid primate, Old World monkey, and chimpanzee before reaching humans—first a Tasmanian and then, at the highest level, a Roman athlete. Gregory observed that some of his paleontologist colleagues seemed to exhibit a "pithecophobia," or dread of apes as relatives or ancestors. His own analysis, like most others of the 1920s and 1930s, assumed the racial and cultural superiority of Caucasians (here represented by a classical bust).

18

*P*hysical *A*nthropology

The evolutionary origins of the human species aroused intense discussion during the century. Debate was inflamed particularly by presumed evolutionary affinities between man and ape. While physical anthropologists focused on the accumulation and evaluation of fossil evidence for human evolution, questions continually arose among the public at large about the possible meaning of such data for understanding the nature of human society.

The earliest twentieth-century discussions of evolution reflected a tension that had arisen between the scientific and popular domains in the nineteenth century. Scientists persistently sought to focus on the biological nature of human life and human capabilities, but the issue was complicated by the cultural differences observed within the human species. Moreover, nineteenth-century scientists had assumed that evolution was a progressive process and that a hierarchy existed within the human family tree. Race, which played a particularly important role in this analysis, was very difficult to evaluate scientifically because of the cultural differences among the human groups. Nonetheless, it was a key concept in most turn-of-the-century evolution discussions.

Scientific treatises often reflected the cultural and racial biases of their authors. In certain countries such writings were used to reinforce prejudice and encourage racial hatred in the decades prior to World War II. In the nineteenth century the racial debate had focused on analysis of

cranial capacity. In the twentieth, new subjects of contention appeared—IQ studies and blood group analysis, the latter thought to be a biological method for firmly distinguishing racial groups. Biological theories were used as a basis for discrimination against Jews and people of color in the first half of the century. Nazi Germany justified a policy of mass extermination with such theories.

In the United States the meaning of fossil evidence was the subject of public controversy and legislative action from early in the century. A memorable trial took place in 1925 when a schoolteacher, John Scopes, challenged a Tennessee law prohibiting the teaching of evolution.

Despite such governmental attempts to regulate the content of scientific discussion, discoveries of new human fossils continued to excite much interest. Java man, the first fossil representative of the early human species now called *Homo erectus*, had been discovered on Java in 1891. A Neanderthal skeleton was found at La Chapelle-aux-Saints, France, in 1908, and other fossil evidence of early man steadily accumulated. A celebrated find on Piltdown Common, Sussex, in 1912 was actually a hoax, but excitement was such that the Piltdown hoax was not exposed until the 1950s.

A fossil older than *Homo erectus* was found in South Africa in 1924 by the Australian-born physical anthropologist Raymond Dart, who dubbed it *Australopithecus africanus* (southern ape from Africa). Particularly rich were fossil finds made in Olduvai Gorge, Tanganyika (since 1964,

Tanzania), East Africa, by the British researchers L. S. B. Leakey and Mary Leakey. These included the 1959 discovery of the early tool-making hominid *Zinjanthropus*, with an estimated age of 1.75 million years, as well as the later discovery of the species called (by L. S. B. Leakey) *Homo habilis*. In 1974 a nearly complete skeleton of the first known upright hominid, *Australopithecus afarensis*, was found in Ethiopia. The skeleton, called Lucy, was thought to be 3.6 million years old.

Meanwhile, in the 1960s researchers began using blood proteins and chromosomes to evaluate the possible genetic links between human beings and apes. By this time, scientists were routinely employing cultural data as well as physical evidence and carbon-dating techniques to place each new skull-type in a branch of the evolutionary tree leading to *Homo sapiens*.

Scientific interest in skin color and brain capacity as indices of evolution gradually gave way to the study of environmental and archaeological evidence that could shed light on the behavior of early humans. At mid-century, human evolution also became much more fully integrated with the general theory of biological evolution. Field studies of the behavior of primates, such as the chimpanzee and gorilla, shed new light on the capabilities of these animals. This work increasingly involved analysis of primate social systems, as well as study of the ability of the animals to communicate through the use of symbols and to devise and use tools. Physical and cultural anthropology, ethology, and ethnology all contributed to a new understanding of the human condition, challenging simplistic theories of biological determinism.

In paleoanthropology, as in paleontology, scientists' theories were limited by the incompleteness of skeletal remains. Consequently, as new finds were revealed, anthropologists debated the precise position of each newly discovered bone fragment on the evolutionary tree. Study of the physical evolution of the human species, however, remained a prominent theme. Scientists tried, particularly after World War II, to eliminate their own social and cultural prejudices in assessing human evolutionary history. A 1964 Unesco-sponsored conference of scientists held in Moscow produced a notable statement on biological aspects of the race question, "Proposals on the Biological Aspects of Race." It reflected a strongly felt belief that "racist theories can in no way pretend to have any scientific foundation and the anthropologists should endeavor to prevent the results of their researches from being used in such a biased way that they would serve nonscientific ends." Such attitudes, resulting from recognition of the genocidal consequences of early twentieth-century racial theories, were reflected in the research and writings of the late twentieth-century biologists, whose work focused more and more on human variability and adaptation to different environments rather than on human differences.

264. Evolution and the public. After World War I a crusade unfolded in the United States against the theory of evolution. For cartoonists, the issue was a godsend. But the consequences were serious. Tennessee in March 1925 made the teaching of evolutionary ideas a crime.

265. Cartoon from the Scopes period. Darwin's "ape-man" theory of evolution found increasing support among scientists in the early years of the century. The eminent paleontologist Henry Fairfield Osborn, however, argued instead for a "dawn-man" theory, with the family of man and the family of apes sharing a common ancestry. There was no conflict over basics: Osborn vigorously defended the notion that humans had evolved. But his defense stressed the concepts of superior and inferior races, and he was an ardent advocate of eugenics.

266. Climax of the Scopes trial. In July 1925 the high school teacher John T. Scopes was tried in Dayton, Tennessee, for violating state law by teaching the theory of evolution. The trial became a dramatic confrontation between two famous lawyers: William Jennings Bryan for the prosecution and Clarence Darrow for the defense. The cross-examination of Bryan (*left*) by Darrow (*right*) was watched by 2,000 spectators.

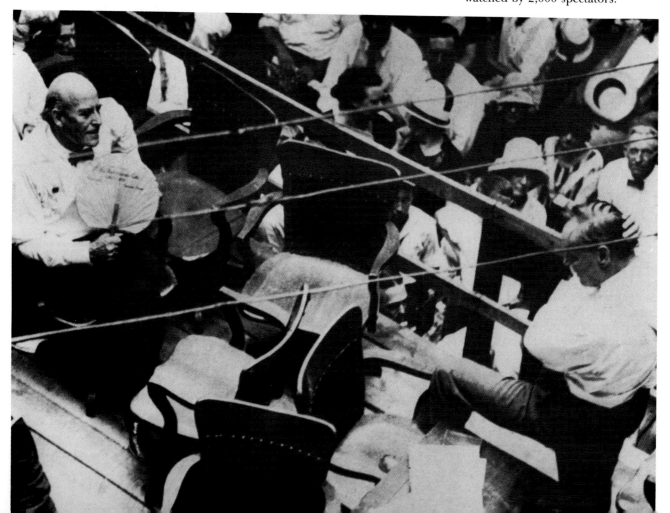

267. Scopes sentenced. Although found guilty, Scopes was fined only $100. Two years later the Tennessee supreme court reversed the judgment on a technicality. The law remained on the books until repealed in 1967. By making Tennessee an object of widespread ridicule, the Scopes trial may have been partly responsible for the fact that few states followed its lead and enacted "monkey laws."

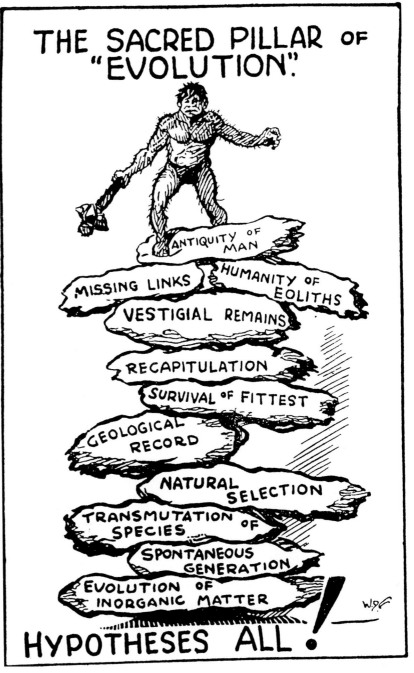

268. Man's evolution from the apes: lack of evidence. This drawing is from an early 1940s book by a "London journalist" (Newman Watts). He was far from the only popular writer to claim that the theory of evolution was built on speculative fantasies and suffered from gaps in supporting evidence. Many commentators feared that science's challenge to the literal truth of the Bible would ultimately undermine Western religious thought, with profoundly disturbing social effects. Sir Ambrose Fleming, a fellow of the Royal Society, said that the book's final chapter might be especially useful "in directing attention to the poisonous effect of evolution theories on the spiritual life of nations."

269. Piltdown man. Portions of a cranium and jawbone that were found on Piltdown Common, Sussex, in 1912 were widely believed to be the long-sought missing link between ape and human. The jaw was apelike, and the skull very much resembled that of modern humans. As fossil evidence of hominids accumulated—with jaws tending to resemble human jaws, and crania simian skulls—the Piltdown man became an anomaly. In the 1950s it was proved a hoax: the skull was from a modern human, the jaw from an orangutan. Among those examining the skull in this 1915 painting are the discoverer, the lawyer and amateur archaeologist Charles Dawson (*rear, second from right*), and Arthur Smith Woodward (*rear, far right*), an anatomist who helped Dawson identify the finds and may have been the intended victim of the hoax.

270. La Chapelle Neanderthal. The Piltdown find was greeted with particular enthusiasm in Britain, perhaps because of a desire to compete with other countries where interesting hominid remains were being unearthed. For example, a notable Neanderthal skeleton was discovered in 1908 at La Chapelle-aux-Saints, France; the head (*below*) was the most fully preserved of any of its kind found up to that time. At right is a comparison of the restored La Chapelle skeleton with the skeleton of an Australian aborigine (*far right*), from the major 1913 monograph on the discovery, by the prominent paleontologist Marcellin Boule. Based on detailed anatomical studies, Boule believed that Neanderthal man was a distinct species and not an ancestor of modern man; his view long dominated French paleoanthropology. A reevaluation of the significance and cultural sophistication of Neanderthal man began in the 1950s.

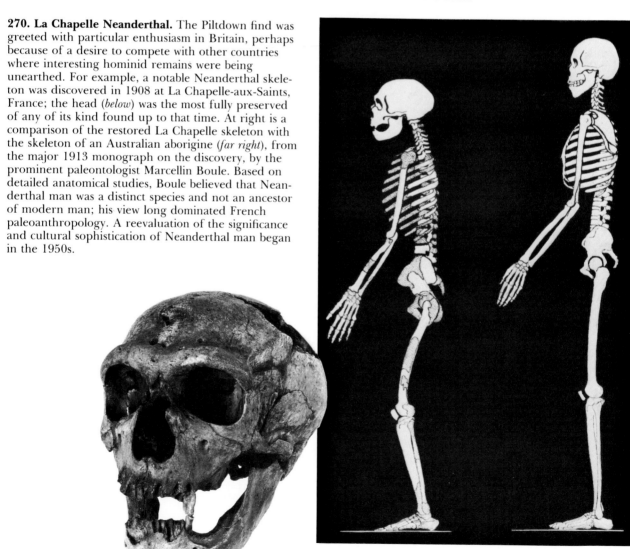

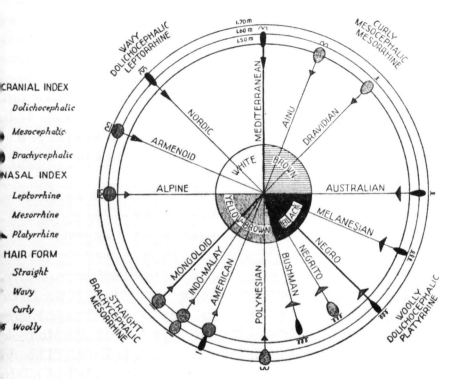

CRANIAL INDEX

Dolichocephalic

Mesocephalic

Brachycephalic

NASAL INDEX

Leptorrhine

Mesorrhine

Platyrrhine

HAIR FORM

Straight

Wavy

Curly

Woolly

271. Racial classification. This diagram from a 1930s work by the American physical anthropologist Earnest Hooton, one of the century's more prominent students of racial classification, exemplifies the classifiers' traditional reliance on physical traits.

272. Trends in hominid brain size. The diagram below is how Henry Fairfield Osborn saw the state of knowledge in 1930. Note the inclusion of Piltdown man. The brain capacity attributed to Piltdown was within the range of modern humans, and the Piltdown cranium seemed very much like modern skulls, yet Piltdown was thought to be very ancient. It seemed then that the human lineage extended extremely far back and that apes, Neanderthal man, and modern humans might represent separate lines of development. This view, of course, accorded with Osborn's dawn-man theory and discredited the ape-man hypothesis linked to Darwin.

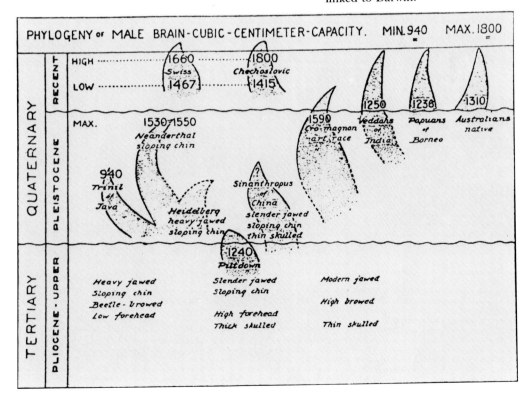

PHYLOGENY of MALE BRAIN-CUBIC-CENTIMETER-CAPACITY. MIN. 940   MAX. 1800

273. Olduvai. The steep-sided ravine in Tanzania called Olduvai Gorge has provided an abundant collection of fossil fauna for researchers. Strata particularly rich in hominids are perhaps 1.75 million years old. The British archaeologist Mary Leakey made a major find in 1959: a large skull of an early australopithecine hominid that was initially dubbed *Zinjanthropus*, or East Africa man. It subsequently became a subject of taxonomic controversy. Other spectacular finds followed.

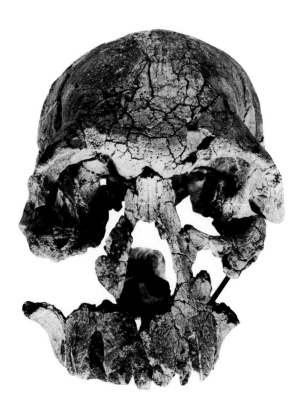

274. Skull KNM-ER-1470. Estimated to be as much as 1.8 million years old, the skull was found at Koobi Fora (formerly East Rudolf) in Kenya in 1972 by a team headed by the paleoanthropologist Richard Leakey (Mary Leakey's son). The brain of the creature was rather large, which suggested that it belonged to the genus *Homo* rather than *Australopithecus*. The jaw and forehead were also humanlike.

275. Lucy. A nearly half complete australopithecine skeleton of a twenty-one-year-old female was found at Hadar, Ethiopia, in 1974. Popularly dubbed Lucy, the skeleton was dated to perhaps 3.6 million years ago. It came from a slender, small-brained individual about 1 to 1.2 meters high who walked upright. Standing next to Lucy is one of her discoverers, the American anthropologist Donald Johanson.

276. Village girls in dance costumes, Samoa. The American anthropologist Margaret Mead, particularly noted for her studies on the psychology and culture of peoples of Oceania, pioneered in the documentation of ethnographic research with photography. This picture is from Mead's first expedition (1925-1926) to the South Pacific. The girl at left has been honored with the title "taupo."

19

*H*uman and *A*nimal *B*ehavior

The methods of measurement that penetrated biological science in the nineteenth century were thought capable of leading to new insights in all areas of research, including the study of behavior. Just as anthropologists adapted the techniques and tools of paleontologists to evaluate the physical nature of man, so twentieth-century zoologists borrowed from the experimental physiologists and psychologists to explain human behavior. Behavioral studies came to reflect basic features of the laboratory work newly characteristic of the life sciences—both in the use of experimental techniques and in the emphasis on measurement and experimental manipulation of variables. This laboratory research was coupled with the traditional observational techniques of the naturalists to construct bold and comprehensive theories asserting the ultimate biological basis for human behavior.

Explaining divergent human cultural traits within a unified, biologically based theory became the goal of biologists, psychologists, ethologists, and ethnologists alike. Over the years their studies ranged from direct introspective analysis of mental events to quantified research on the recurring features of human and animal behavior. Sigmund Freud, who trained as a physician, sought to create a distinctive theory of the mind, which he tried to superimpose on the observable physical characteristics of the brain. Less concerned with localizing functions in specific areas of the brain than his contemporaries, he created a new theoretical framework for understanding human behavior. However, many biologists rejected his approach as unscientific. It did not apply the quantitative methodology espoused in the other rising life sciences.

In the first part of the century many psychologists sought to apply the fundamental scientific criteria of objectivity and quantification to the study of mental and behavioral events. This was especially evident in the mental testing and IQ studies undertaken in the early decades. The result was a tension between laboratory studies and clinical observations that persisted for the rest of the century. When challenged in the early decades by the new psychoanalytic view of the mind, the methodological commitments to objectivity and quantification produced a multitude of schools of experimental psychology, each emphasizing a particular research technique or theoretical construct.

Direct observation of the behavior and characteristics of animals, especially primates, seemed to many researchers to provide a more reliable framework for study of the origin of human emotional and intellectual traits than did psychoanalytic theory. Such research took its cue from Darwin's evolutionary speculations and from the quantitative techniques emerging as central in laboratory studies and field observations. The behavior of animals provided reproducible data with which to explore the nature of instinct and its relationship to learned behavior in humans. For example, the Austrian zoologist Konrad Lorenz, a pioneer in ethology, examined the behavior of birds, especially geese. Other research—including that of fish behavior—probed further into the relationship between genetically determined and learned behavior.

Biology's increasing emphasis on experimental manipulation in the early years of the century had an important impact on psychological research, culminating in the work of the American behavioral psychologist B. F. Skinner. The "Skinner box," for the study of operant conditioning in animals, became a well-known piece of equipment in experimental psychology laboratories in the middle decades. Experimental techniques were also used to isolate specific behavior patterns and to study their biological determinants. For example, the American psychologist Harry Harlow studied mother-infant bonding in rhesus monkeys by substituting a wire mesh mechanical surrogate for the natural mother.

As the years passed, the biology of social organization was studied more and more by experimental techniques. Researchers created artificial environments in which small animals—rats, say, or mice—pursued their daily lives. Such laboratory-simulated field studies enhanced understanding of the behavioral effects of overcrowding by focusing on how mammals adapt to changes in their environment.

Some scientists eventually used social organization among lower animals to comment on human organization and behavior. The behavior of social insects like ants was interpreted to explain the biological origins of such important political concepts as slavery, subordination, and dominance within and among species. The viewpoint and data of sociobiology, as this approach was called, were hotly debated in the mid-1970s. Left-wing and feminist critics challenged the underlying assumptions of sociobiology. They asserted that definitions of what is natural in human behavior are largely culture bound and linked to the established social order. Thus, it was claimed, that which is termed "natural" in behavior may not necessarily reflect genetically determined characteristics.

Despite biology's increasingly genetic view of the organism, toward the end of the century efforts to isolate the human animal from its environment and to study "natural" human properties were judged largely unsuccessful. Human study of biological processes had transformed the earth, altering the occurrence of pathological organisms, raising the productivity of cultivated crops, and accelerating demographic shifts and changes in the earth's topography and resources. Through its socioeconomic consequences, biological research had had a profound impact on human perceptions of the natural world. These effects had in turn altered human experience and behavior and influenced later twentieth-century expectations of human progress and of scientific research. Traditional concepts of human rationality were undermined when scientists advanced contrary, but seemingly equally convincing, explanations of the same data. Particularly in IQ and gender studies, the fact that alternative interpretations could be constructed showed that to a large extent interpretation remained culture bound. This raised fundamental questions about the purported disinterestedness of biological research. The goals of research areas like IQ and gender studies came under further scrutiny as their political and economic consequences were laid bare.

277. The Psychiatrist. This turn-of-the-century caricature mocks what was regarded as the unscientific approach of those who treated mental problems. The doctor, whose room is curiously well supplied with skulls, remarks: "When someone comes to me for examination, I have him count. If he counts well, it's dementia praecox; if he counts badly, dementia senilia."

278. Freud's consulting room. The revolution wrought by Sigmund Freud in psychiatry beginning in the closing years of the nineteenth century had a far-reaching influence on Western culture. He put forth a controversial, yet comprehensive theory of the human mind and developed the technique of psychoanalysis for treating neuroses. In the consulting room of his Vienna flat Freud analyzed the dreams and memories of his patients. He sat in the corner chair behind the head of the couch, on which the patient lay. Freud believed that the patient, not seeing him, felt less inhibited and free-associated more readily. He spent exactly fifty-five minutes with each patient.

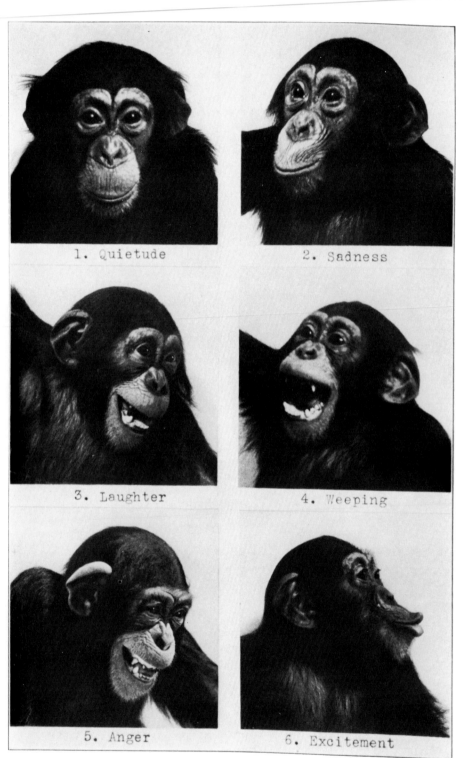

1. Quietude   2. Sadness

3. Laughter   4. Weeping

5. Anger   6. Excitement

279. Chimpanzee facial expression.
These photographs of young chimpanzees expressing emotion were among the best to be produced in the first decades of the century. They were made in the laboratory of the prominent Soviet animal psychologist Nadezhda Ladygina-Kots (Kohts) at the Darwin Museum in Moscow. The century saw photographs increasingly supplement descriptive accounts of the emotional life of animals, as scientists attempted to exclude subjective and anthropomorphic influences from their work.

280. Gorilla using symbols. In a project beginning in the 1970s, the Stanford University researcher Francine Patterson taught the female gorilla Koko to use sign language, with a vocabulary of several hundred signals. Gorillas lack the vocal apparatus for human speech, but Koko learned to produce spoken words by using a computer-linked auditory keyboard. Here, Koko signs "Smoky" for her pet cat.

281. Goodall feeding chimpanzees. From 1960 the British ethologist Jane Goodall carried out a long-term program of observation of free-ranging chimpanzees in their natural habitat in the Gombe Stream Game Reserve, Tanzania. The project, with techniques refined as the years passed, eventually became the longest continuous field study of any animal. Goodall's findings included the unexpected: the occurrence of murder and cannibalism—thought by some to be due to feeding of the animals and increased population densities. Goodall also obtained valuable new data on intergroup aggression, kinship bonding, and cognitive abilities.

282. Triumph ceremony. In studying animal behavior patterns the Austrian zoologist Konrad Lorenz and his associates carried out close comparative analyses of gestures and movements, paying attention to their motivations and evolution. Here we see the motor patterns of part of the "triumph" rite performed by geese. These gestures, which bond families or groups of individuals together, were described in Lorenz's popular work *On Aggression* (1966). Lorenz used such data to suggest an inborn basis for fighting and warlike behavior in humans.

283. Lorenz followed by geese. Lorenz was known particularly for his work with birds. Beginning after World War I he played a key role in transforming the study of animal behavior into the science called ethology, focusing on both instinctive and learned ("imprinted") behavioral acts and on the evolution of behavior patterns. He also pioneered in the study of animal communication. His methods emphasized direct observation and analysis of animal behavior rather than the manipulation of animals in experimental settings favored by the behavioral school.

285. Starling at feeding machine. Rats were certainly not the only animals used in research on rewards and punishments in learning. This experimental setup was typical of many mid-century bird activities studies. The animal learns to acquire food by operating on its environment; reinforcement is controlled in part by the animal's own behavior, in part by the research plan.

284. Skinner. Here B. F. Skinner, the leading American behavioral psychologist, works with a rat in an operant conditioning chamber. The rat's "responses" (such as pressing the horizontal bar) are reinforced with rewards of food. The recorder keeps track of the responses.

286. Wellsprings of attachment. In a 1950s study by the American psychologist Harry Harlow, infant rhesus monkeys were placed with two mechanical mother-surrogates. One was of wire mesh; the other was covered with soft cloth, to which the infants could comfortably cling. The monkeys preferred the cloth mother, even when a nursing bottle with milk was attached to the wire mother. Harlow concluded that contact comfort is a major factor in building the mother-infant bond.

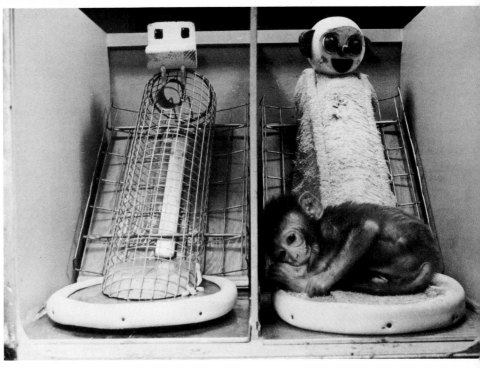

287. Two ant species: slaves and slave makers. Sociobiology, which focuses on the role of genetics and natural selection in shaping animal and human social organization, entered the arena of public debate in the 1970s. "The similarity of the slave maker to the enslaved ant," wrote the American sociobiologist Edward Wilson, "is an example of the rule of phylogenetic closeness that applies to most kinds of social parasitism."

288. First simulated rat universe. In experiments beginning about mid-century, the American ecologist John Calhoun studied patterns of social organization and population dynamics among such small mammals as rats and mice. His first rat "universe," 100 feet square, was built in a field behind a house in Towson, Maryland. An outside wall of the universe is visible in the background. Calhoun can be seen in the foreground taking notes on the characteristics of a rat burrow. Food and water were located in the central pen, which was 20 feet square.

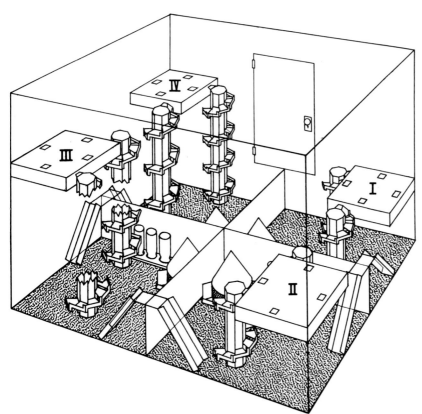

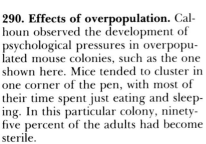

289. **Four-cell rat universe.** In this experimental setup Calhoun discovered what he called the behavioral sink phenomenon. Each cell had a raised "apartment house" reachable by spiral stairs, as well as a conical food hopper and water bottle. There were ramps over the partitions between the cells, except between cells I and IV. The rats tended to favor cell II because its "apartment house" was at a low height and because II was the cell with the most access to the other cells. Calhoun found that rats not only became accustomed to eating together but that this became a felt necessity. Most of the rats came to live in cell II, even though the population density there became pathological.

290. **Effects of overpopulation.** Calhoun observed the development of psychological pressures in overpopulated mouse colonies, such as the one shown here. Mice tended to cluster in one corner of the pen, with most of their time spent just eating and sleeping. In this particular colony, ninety-five percent of the adults had become sterile.

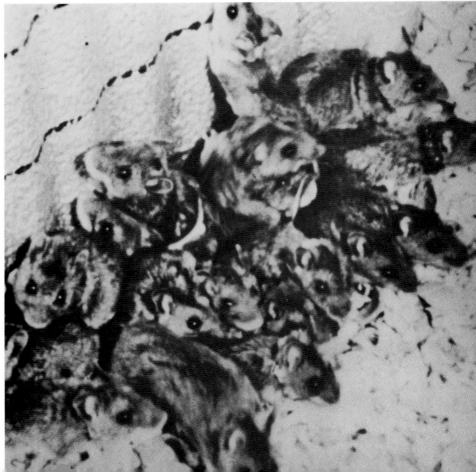

Part Six

APPLYING
BIOLOGY TO
SOCIAL NEEDS

291. Ehrlich. During a research career bridging the nineteenth and twentieth centuries, the versatile German medical scientist Paul Ehrlich played a seminal role in the development of hematology and immunology and discovered a specific treatment for syphilis—the compound Salvarsan (arsphenamine). Ehrlich introduced the term "chemotherapy" for treatment with synthetic drugs that selectively attack disease-causing organisms.

CHAPTER

20

Controlling Disease

The development of antibiotics was one of the most stunning achievements of twentieth-century laboratory research. These substances were an extension of the class of medications now called biologicals (since they are produced by biological organisms). The biologicals and artificially produced drugs introduced during the century helped fulfill an age-old hope of Western medicine: the discovery of specific drugs that could cure specific diseases. The new drugs—the products of living organisms as well as the artificially synthesized chemicals—became part of every physician's armamentarium in the second quarter of the century. The isolation and manufacture of such substances required the combined efforts of microbiologists, pharmaceutical chemists, and industry. The causative organisms of disease first had to be identified; drugs that inhibited or killed them then had to be discovered, isolated, characterized, and produced commercially on a large enough scale to have a significant impact on the incidence of disease and the associated death toll.

The first malady to be attacked fruitfully by new artificially produced chemicals was the dreaded venereal disease syphilis. Early in the century the German medical scientist Paul Ehrlich discovered what he called substance 606 or Salvarsan, later known by the generic name arsphenamine. He used it as a "magic bullet" to attack the treponema bacteria responsible for syphilis. Ehrlich introduced the term "chemotherapy" for the use of synthetic chemicals capable of inactivating pathogenic organisms.

The success of Ehrlich's research program was closely connected with the rise of the German chemical industry and was based on earlier work by the bacteriologists Robert Koch and Emil von Behring. In the late nineteenth century coal-tar derivatives had yielded new histological stains allowing microscopists to observe more clearly both bacteria and the body's response to infection. Between 1880 and 1900 the causative organisms for most of the major infectious diseases had been identified. New drugs had been introduced, including the popular painkiller aspirin. Meanwhile, German and French researchers had developed specific antitoxic sera against diphtheria and tetanus. Expectations of laboratory medicine in those years were high, fed by the emergence of two new therapeutic movements based on the chemical products of living organisms. In addition to serotherapy—treatment with blood serum from immune animals—there was organotherapy, treatment with organ extracts.

Nonetheless, at the turn of the century prevention was still generally a much more successful strategy for fighting disease than were curative efforts. Infectious disease was declining, but this was due primarily to improvements in sanitation, housing, hygiene, and nutrition rather than to the use of drugs or to other direct medical intervention. Adherents of the new scientific medicine, however, were convinced that breakthroughs in curative medicine were close at hand. Some time would pass, but they would be proved right.

In the first decades of the new century medical scientists continued to search for chemical cures. A major breakthrough came in the 1930s. Gerhard Domagk, working at a research laboratory in a German dye factory, discovered that a dye called prontosil was effective against streptococci. In 1935 researchers at the Pasteur Institute in Paris reported that, once in the body, prontosil released its active component, which was sulfanilamide. In 1938 an analogue, sulfapyridine, was found effective against pneumococcal diseases. These "sulfonamides," or sulfa drugs, opened the door to physicians' reliance on chemotherapy.

It was the discovery of penicillin, however, that eventually revolutionized the medical treatment of disease. In 1928 the British bacteriologist Alexander Fleming observed an area free of bacteria around a mold growth in a laboratory dish. The antibiotic effect of a substance produced by the mold, *Penicillium notatum*, was found to occur with bacterial cultures of staphylococci, pneumococci, and streptococci. In order for the discovery to have an impact on clinical practice, however, the active agent had to be purified and produced on a large scale. For this achievement the Australian pathologist Howard Florey and the biochemist Ernst Chain, a German refugee, shared the 1945 Nobel Prize in physiology or medicine, together with Fleming, who was honored for his discovery.

At Oxford University in the 1930s, Chain and Florey developed production and extraction techniques to provide penicillin for clinical use. Then, during World War II, Florey worked with the American Committee on Medical Research and three pharmaceutical companies on ways to make the drug in large quantities. The technique of deep-tank fermentation was developed to increase output. During the final two years of the war penicillin was used to treat war wounds. Researchers soon found it effective against syphilis as well. Thus, biologically synthesized substances were now competing vigorously with artificially synthesized compounds in the treatment of infectious diseases. Physicians began to observe, however, the development of microorganism strains resistant to the new medications—a problem that would later cause them to be more circumspect about routinely prescribing the drugs.

Large-scale industrial production of antibiotics flourished after World War II. As the pharmaceutical industry expanded, research intensified. Streptomycin proved effective in the treatment of tuberculosis and tularemia. Other antibiotics followed, including chloramphenicol, the tetracylines, and the cephalosporins. These "miracle drugs" enhanced the physician's power and social status and encouraged the public to believe that laboratory research would produce a medical science with unlimited curative powers. Treatment of disease with drugs, rather than prevention, was the banner of the postwar era, as hope spread that all disease would succumb to scientific medicine.

In later decades, however, preventive medicine slowly regained favor, although it was generally less lucrative and less glamorous a calling than the laboratory-based medicine characterizing the first half of the century. The preventionists reasserted the claim that disease could be avoided by relatively simple and inexpensive hygienic, nutritional, and environmental measures.

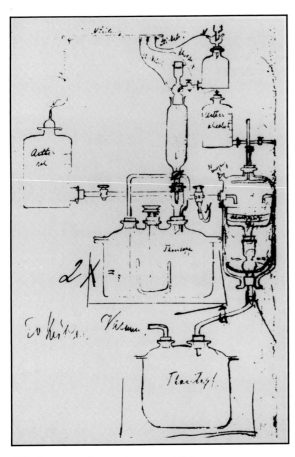

292. Large-scale production of Salvarsan. Ehrlich's research program included the synthesis and testing of many related compounds. Over 600 arsenic compounds were tested for therapeutic effect. Salvarsan (from the Latin *salvare*, "to preserve," and *sanitas*, "health") was number 606. This freehand drawing by Ehrlich outlines apparatus for Salvarsan production. Ehrlich was an early advocate of cooperative effort between research and industry. Such links were later strengthened by the discovery and production of sulfonamides and penicillin.

293. Testing for syphilis. In a Chicago street scene from before World War II, a physician takes a blood sample. Strapped to the back of the physician's assistant is a case with test tubes and equipment. The serum test for syphilis was developed in 1906 by August von Wassermann, Albert Neisser, and Carl Bruck. It rapidly became the standard laboratory test for the disease. The testing shown here was supported by the Works Progress Administration during the Great Depression.

294. Domagk. In the 1930s the German bacteriologist and pathologist Gerhard Domagk was studying dyes for possible antibacterial effects when he discovered that the dye prontosil could indeed combat strepto-coccal infections. Prontosil and its active part, sulfanil-amide, were the first of a class of chemical substances inhibiting bacterial growth that became known as sulfa drugs. These "sulfonamides" were greeted as wonder drugs; they sharply cut the number of deaths from such bacterial diseases as pneumonia and blood poi-soning and saved many lives during World War II. Domagk was awarded the Nobel Prize in physiology or medicine in 1939.

295. Sulfa drug in action. Below left are chains of pus-discharging strepto-coccus bacteria. Below right, a sulfon-amide has attacked the streptococcal chains.

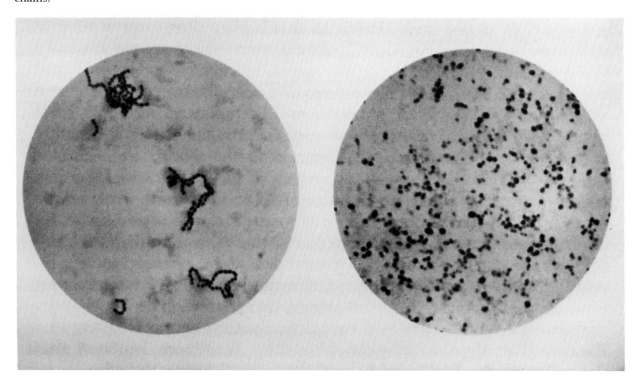

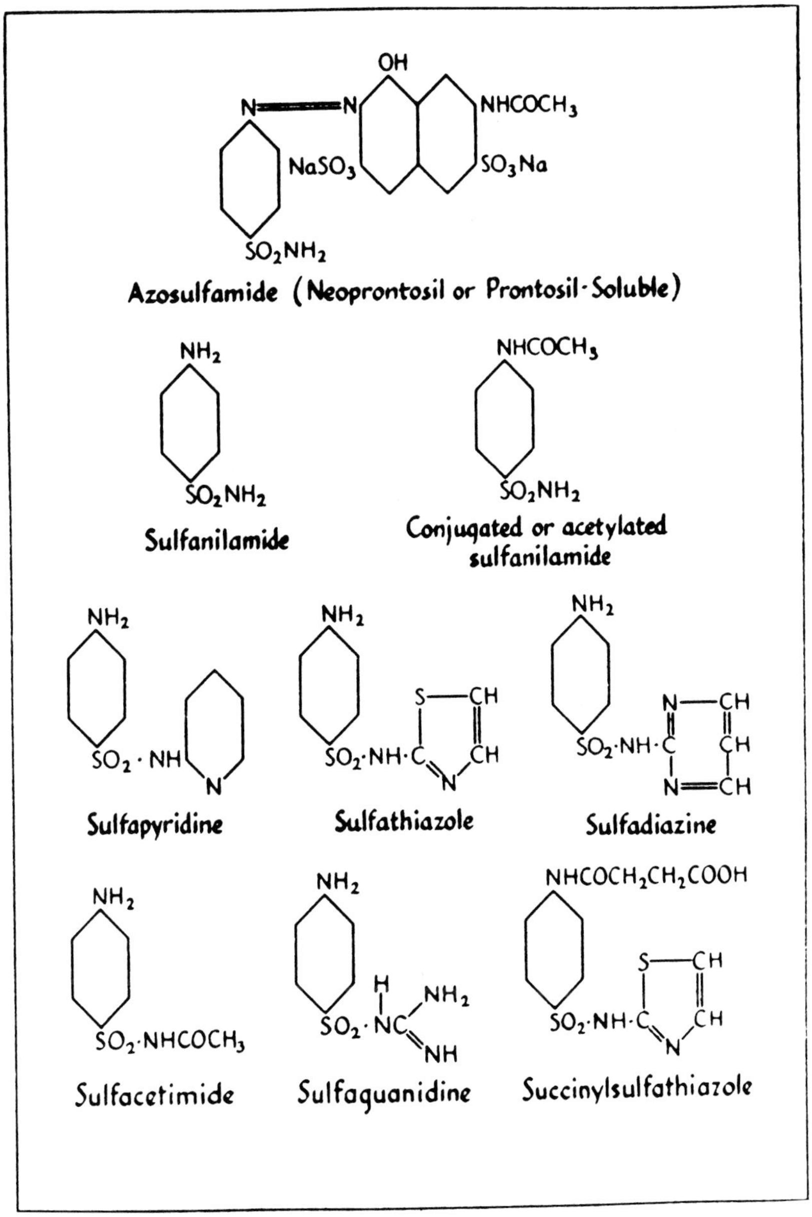

OH

N══N ⬡ NHCOCH₃

NaSO₃   SO₃Na

SO₂NH₂

Azosulfamide (Neoprontosil or Prontosil·Soluble)

NH₂

SO₂NH₂

Sulfanilamide

NHCOCH₃

SO₂NH₂

Conjugated or acetylated sulfanilamide

NH₂

SO₂·NH⬡N

Sulfapyridine

NH₂

SO₂·NH·C

S——CH
‖   ‖
N   CH

Sulfathiazole

NH₂

SO₂·NH·C

N——CH
‖   ‖
       CH
N══CH

Sulfadiazine

NH₂

SO₂·NHCOCH₃

Sulfacetimide

NH₂

SO₂·NC
H

NH₂
|
C
‖
NH

Sulfaguanidine

NHCOCH₂CH₂COOH

SO₂·NH·C

S——CH
‖   ‖
N   CH

Succinylsulfathiazole

296. Chemical structure of sulfonamides. A number of sulfonamides were synthesized in the late 1930s. These diagrams depict the chemical relationships between some of the drugs and underscore the important role played by organic chemists in the production and analysis of the new chemotherapeutic agents. Sulfonamides alter the metabolism of microorganisms and thus prevent their multiplication. Bacteria, however, can develop resistance to the drugs.

297. Discovery of penicillin. In the center of the plate of agar are three colonies of the mold *Penicillium notatum*. Bacterial growth is evident at the periphery, but in the dark areas around the colonies the bacteria have been inhibited by the mold. This is how the Scottish bacteriologist Alexander Fleming discovered penicillin in 1928, thereby opening the era of antibiotics—drugs even more potent than sulfonamides against bacteria. In Fleming's case, a culture of staphylococci was contaminated accidentally by mold organisms in the air. Fleming was able to derive an unstable crude kind of penicillin from the mold culture and found it effective against several kinds of bacteria. Practical application of Fleming's discovery was not possible until after Ernst Chain and Howard Florey succeeded in purifying penicillin in 1940.

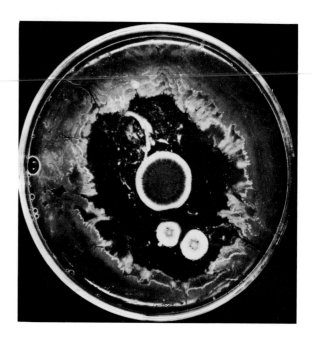

298. Guises of penicillin. The flasks above contain penicillin mold; here the mold grows abundantly, releasing penicillin into the medium. Chain and Florey developed methods for concentrating and isolating the drug so that it could be used for therapeutic purposes. At right we see penicillin in crystalline form. The isolation of pure crystals of the drug permitted chemical analysis. The mold produced several related organic acids. A small amount of the sodium salt of penicillin G (benzyl penicillin) was established as the international unit of potency at a conference of the League of Nations Health Organization held in 1944.

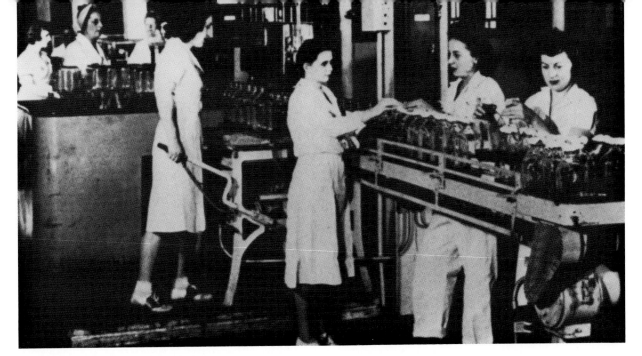

299. World War II penicillin production. Initially penicillin was obtained by growing the mold *Penicillium notatum* on the surface of a culture medium in a small container. With wartime urgency the U.S. government pushed for large-scale penicillin production, and assembly lines appeared. The workers in this 1944 scene are sterilizing half-gallon milk bottles, adding nutrient, and sealing the bottles. The bottles would next be placed on their sides in racks and inoculated with mold spores.

300. Fleming inspects a deep tank. A major step forward in the U.S. push for mass production of penicillin was the development of "submerged fermentation." Growing the mold in the nutrient medium rather than on top of it meant, eventually, the introduction of large "deep tank" production units. Other technological problems had to be solved as well, and were. Between 1943 and 1945 more than twenty penicillin plants were constructed in the United States. By the end of the war, production had reached 650 billion penicillin units a month. The price stood at twenty dollars a dose in July 1943, and fifty-five cents three years later. The deep tank in the photograph, taken during a visit by Fleming to a New Jersey plant, has a capacity of 15,000 gallons.

301. Streptomycin plant. Here is a late 1940s view of the part of a Pfizer plant in Brooklyn where the antibiotic streptomycin was recovered from the fermentation liquor. The scaling up of antibiotic production techniques fostered a new specialty called biochemical engineering.

302. Antibiotics research. This investigator is looking for antibiotic agents effective against cancer. The purposeful creation of new drugs and the close association of such research with industry have been major themes in twentieth-century biological science. In the latter part of the century genetic engineering techniques further encouraged this association by providing yet another mechanism for large-scale production of biologicals.

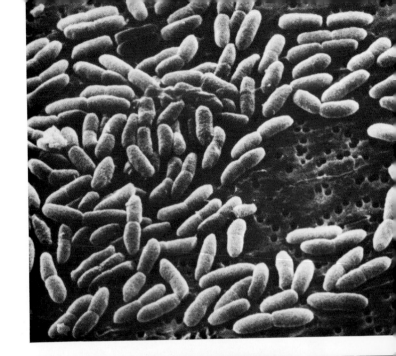

303. Destruction of bacteria by an antibiotic. The bacteria are of the genus *Pseudomonas*, capable of causing infections in wounds and burns. In the top electron micrograph we see them multiplying in a laboratory dish containing a nutrient medium. Some individuals are ready to divide into two new bacteria. The second picture shows the bacteria three hours after the antibiotic moxalactam has been added. As a bacterium grows, moxalactam keeps the cell walls from pinching inward to make two new cells. Thus the bacteria have become extremely elongated: unable to divide, they continue to grow, while their walls deteriorate. In the wrinkled areas a cell wall has ruptured, with cell contents leaking out. The bottom picture was taken six hours after the administration of moxalactam. Nearly all the contents of the bacteria have spilled out, leaving only ghosts, the cell walls.

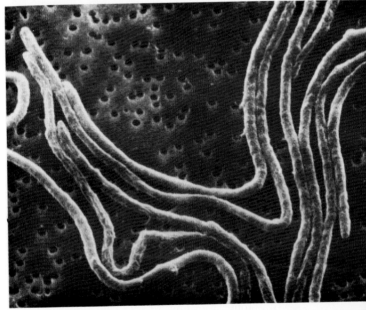

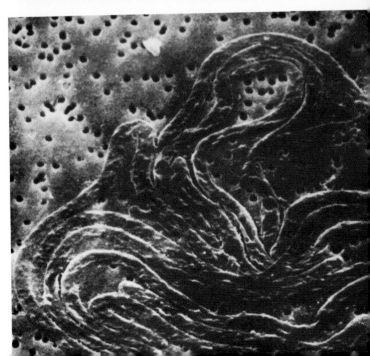

304. The war on smoking. "Don't smoke!" enjoins this Soviet poster from the 1960s. Threatening the smoker, it claims, are disease of the vessels of the brain and heart, endarteritis, and cancer of the lungs, larynx, and lower lip.

305. Cancer control. The billboard reminds the public of the "three earlies" of controlling cancer: detection, diagnosis, and treatment. Specific preventive measures in the Chinese cancer information campaign included a recommendation to dry (rather than pickle) vegetables, in order to eliminate a mold linked to cancer.

CHAPTER

21

*P*reventing *D*isease

The striking successes scored by antibiotics against
infectious diseases during and after World War II
resulted in a temporary waning of public interest in
disease prevention. Nonetheless, preventive medicine continued to make substantial strides. New
vaccines were introduced, and international campaigns were mounted against major killers. Governments took steps to make the environment less
favorable for disease-causing organisms and to
educate people in preventive measures. All in all,
in the course of the century chronic illnesses like
heart disease and cancer replaced infectious diseases as the major causes of death in developed
nations. Many historically important infectious
diseases were nearly (or even completely) eradicated, making it possible for public health advocates to turn their attention to new concerns, like
proper nutrition and a healthful life-style.

Understanding what happens when a person
has, or acquires, immunity to a disease was a major
goal for medical researchers. Building on the work
of Pasteur and Koch, scientists began analyzing the
body's response to microbial infection in the late
nineteenth century. Élie Metchnikoff and Paul
Ehrlich threw light on the immune response's
cellular aspect (the protective activity of white
blood cells) and humoral aspect (antibodies and
other chemical components of blood sera). In the
new century physicians and biologists studied not
only humoral and cell-mediated immunity but also
how disease spreads through a community. Improvements in housing and diet, at least in the

more prosperous countries, had enhanced individuals' ability to resist disease. As the century began,
hopes were high that new vaccines could be
produced imparting resistance to specific diseases.
Physicians already possessed experience with
smallpox immunization and had seen the success
of the newly developed antitoxic sera against
diphtheria and tetanus.

Progress toward new vaccines was sometimes
slow in coming, but other preventive measures
could be refined—or revised—as knowledge accumulated. The preventive treatment for polio, for
example, changed markedly during the century. A
measure used in the 1930s was to spray zinc sulfate
or other chemicals into the nose to guard against
infection. But this tactic was unsuccessful, since
the poliovirus does not enter the body by that
route—it comes in through the alimentary tract.
Researchers eventually developed vaccines against
polio. A killed-virus vaccine was introduced in the
1950s, and a live-virus vaccine in the 1960s. Mass
vaccination programs nearly eradicated the disease
in some countries.

Preventive medicine's most impressive victory
came against smallpox, an ancient affliction of humankind. A worldwide effort was begun in 1967 to
control the disease through vaccination and quarantine. The strategy was to disrupt transmission of
the smallpox virus. This and other coordinated vaccination programs required the cooperation of
many levels of government.

Insect-borne diseases like malaria and yellow

fever proved much harder to eradicate because of the difficulty of destroying the insect carriers. Scientists had studied these diseases intensively from the latter part of the nineteenth century, but further biological research was essential. Partial control was achieved after World War II with the help of insecticides. Insect strains resistant to the insecticides eventually developed, however. Also, objections were raised against some insecticides because they turned out to be toxic to humans or wildlife. In the case of malaria control, efforts focused on elimination of the mosquito carrying the parasite responsible for the disease and on the use of antimalarial drugs. In the case of yellow fever, mosquito control was coupled with the use of an effective vaccine, developed in the 1930s.

In each effective prevention campaign government support played an essential role, showing that success of a disease prevention program depends on participation by the civil authorities. A coordinating role was played by philanthropic agencies like the Rockefeller Foundation (especially before World War II) and, after 1945, by the United Nations and World Health Organization.

Implementation of advances in preventive medicine depends on a variety of social, political, and even economic factors. In the nineteenth century improved hygiene and sanitation were the most important means of lowering the incidence of disease, but in the mid-twentieth century the central concerns of public health officials embraced the distribution of vaccines mass-produced by the pharmaceutical industry and the provision of adequate nutrition to all the population of the world. The political, as opposed to biological, determinants of public health thus received increasing attention. Although scientific medicine provided a way of controlling the major infectious diseases, social factors dictated whether or not that knowledge or technology had a significant effect on a given population.

Preventing the massive incidence of illness and death caused by famine particularly required international cooperation. The situation was complex, because famines could result from political causes (such as war or the enforced collectivization of agriculture in the Soviet Union in the 1930s) as well as from natural causes, like drought and crop disease. To be effective, a relief program required the cooperation of the government of the afflicted nation. The century did see, however, progress in combating famine. Advances in preventive public health in this case included improved agricultural technology for avoiding crop failure and improved facilities for transporting foodstuffs.

The incidence and distribution of cancer, heart disease, and other chronic diseases emerged as a major public health concern in the industrialized world in the second half of the century. In the final decades advocates of preventive medicine found a new focus for their efforts: the maintenance of good health, of "wellness." If miracle drugs had once been the leitmotif of popular health discussions, people now talked about the importance of the individual's own efforts, emphasizing guidelines for eating and living right and promoting prevention over cure as the proper goal of medicine in the twenty-first century.

306. The ravages of famine. Women and children await emergency food at Bume, Ethiopia. The second half of the century saw the distribution of world food supplies come to be viewed as a major responsibility of government authorities and a major focus for public health action. Famine remained a major cause of disease and death in some areas of the globe. In large parts of Africa in the 1980s, for example, severe drought brought millions face-to-face with starvation.

307. Course of an epidemic. This hypothetical disease has an incubation period of one week. Once infected, a person is assumed to become immune. The rectangles at the bottom show the ratio of susceptible (white) and immune (hatched) individuals for each four-week period. Each circle is a case of infection. Black circles are infected persons who do not infect others. Few lines of infection can be continued in the final period, since most individuals have become immune. The diagram, by the Australian immunologist Macfarlane Burnet, is based on knowledge at mid-century of the immune response.

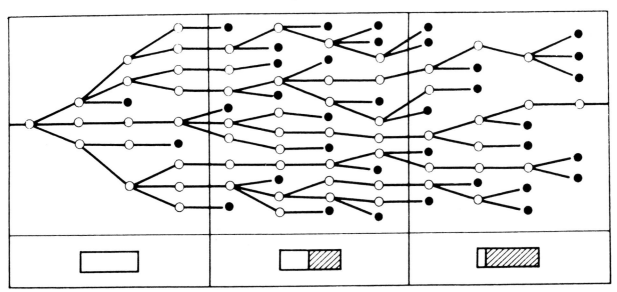

308. Yellow fever research: mosquito collection box. Efforts to combat yellow fever during the century have focused on controlling the mosquitoes that carry the disease and on developing a vaccine—which was successfully done in 1936 by the South African-born American microbiologist Max Theiler. The photograph was taken in 1946 in Ilaro, Nigeria. Inside the mosquito collection box is Rockefeller Foundation researcher John C. Bugher, who established the Yellow Fever Research Institute in Lagos, Nigeria, in 1943 and served as its director until 1948.

309. Yellow fever immunity survey. At right is a 1946 picture from the work of the Yellow Fever Research Institute in Nigeria. Here the researchers are taking blood samples for analysis.

310. Yellow fever animal studies. Below, again a 1946 scene from the Yellow Fever Research Institute, a distillate of ground mosquitoes is injected into the brains of anesthetized mice.

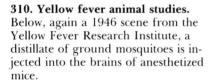

311. The fight against malaria. In the mid-1950s the World Health Organization launched an international campaign to eradicate malaria. Measures included spraying dwellings with insecticide (usually DDT) to kill the mosquitoes that carry the malaria-causing parasites and treating people with antimalarial drugs to kill the parasites in the bloodstream. By the late 1960s eradication had been achieved in most parts of the world outside the tropics and subtropics. Initial victories in many tropical and subtropical areas, however, were followed by a resurgence of the disease. The poster is from a large-scale antimalaria campaign launched in India in the 1960s.

312. Mosquito spraying. A standard tack in the World Health Organization's antimalaria campaign was spraying dwelling places twice a year with a residual insecticide. After a substantial decline in the incidence of the disease, surveillance replaced spraying, except in areas where substantial numbers of cases remained.

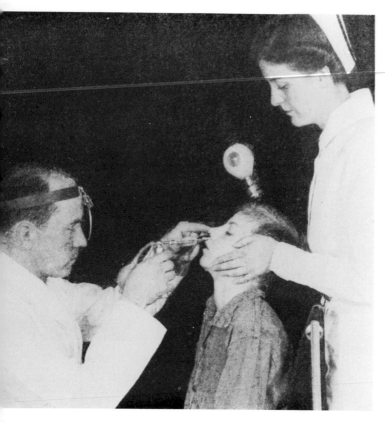

313. Polio prevention. Poliomyelitis, or infantile paralysis, became a serious problem in developed countries in the first half of the century. In the United States a particularly severe epidemic peaked in 1952. The disease was finally brought under control with vaccines. Earlier preventive measures included the approach shown in this 1938 picture. The physician is spraying the nose of a young boy with zinc sulfate, the rationale being that the germs, it was thought, entered the body "by way of the nerves of smell." This hypothesis was later disproved: entry was through the digestive tract. In developing countries people usually acquired immunity early in life through exposure to the virus.

314. Salk testing vaccine. The first safe and effective polio vaccine was developed by the American virologist Jonas Salk in the 1950s. Administered by injection, the vaccine used viruses that had been killed. Here we see Salk giving a shot of his vaccine to a boy during a 1958 mass immunization trial in Pittsburgh.

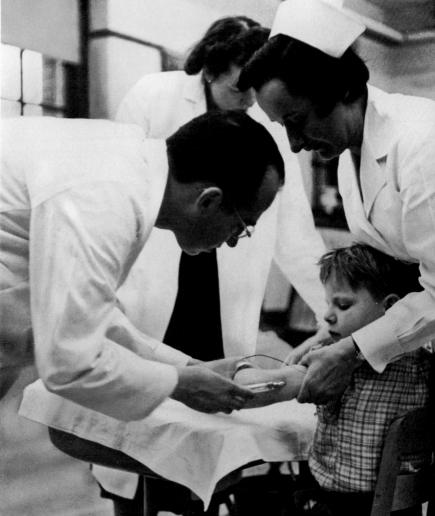

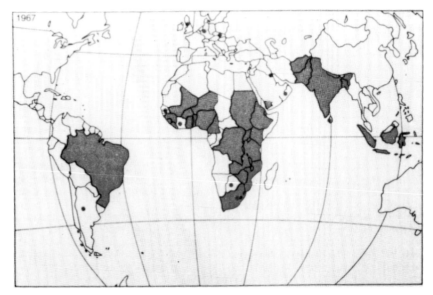

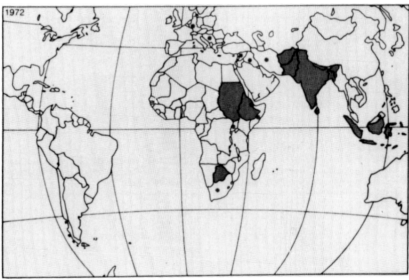

315. Last smallpox victim. In 1967 the World Health Organization began a worldwide program to eradicate smallpox through vaccination and quarantine. The campaign against the ancient scourge of humankind was a success. The last person reported to contract naturally occurring smallpox was the Somalian Ali Maow Maalin, who developed smallpox rash in October 1977. Laboratories where the virus was kept for study remained a potential problem, however, and cases of infection have occurred as a result of faulty containment procedures.

316. Conquest of smallpox. These maps of the occurrence of smallpox in 1967, 1972, and 1975 show the progress made by the World Health Organization campaign. A similar map reflecting the situation as of the end of 1977 would have no shaded countries—no cases of the disease were being reported. After necessary confirmation procedures, WHO announced in May 1980 that naturally occurring smallpox had been eradicated.

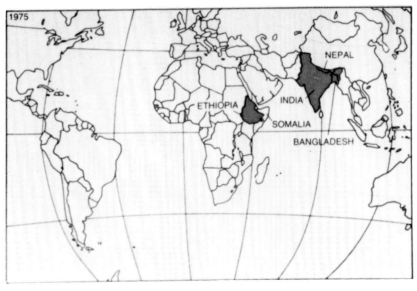

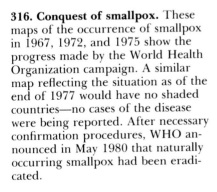

317. A Stopes mobile clinic, early 1920s. The British paleobotanist Marie Stopes, the first woman on Manchester University's science faculty, became a vigorous advocate of birth control during World War I. She initially viewed it as an aid to marital harmony and a means of eliminating the heavy physical strain exerted on women by excessive childbearing; at the time other advocates of birth control saw it primarily as a means of social reform—a way of fighting overpopulation and poverty. By the 1920s eugenic arguments came to the fore. Stopes founded the first instructional clinic for contraception in the United Kingdom (the Mothers' Clinic for Constructive Birth Control) in 1921 and the Society for Constructive Birth Control and Racial Progress the following year. After World War II she was an active proponent of birth control in the Far East.

22

Controlling Population Growth

By mid-century, population control had become an important international issue. Campaigns to cure and to prevent disease had experienced striking successes, leading to a marked and economically threatening increase in the world's population. Programs to deal with the problem focused on providing contraception to all fertile individuals. Early in the century, however, government and the public at large expressly opposed family planning.

The population control issue had a lengthy, controversial prehistory in debates over birth control. Although discussion of the issue in the latter part of the century was fueled by scientific research on the human reproductive cycle that emerged in the middle decades, concerted efforts to control human fertility had begun decades before. The early efforts were a response to: (1) the acceptance, rather than repression, of human sexuality that was evident already in the late nineteenth century and (2) new biological theories about genetic fitness and human evolution.

In the first two decades of the twentieth century many birth control advocates emphasized the connection between poverty and overpopulation, relying heavily on the traditional economic arguments put forth by Thomas Malthus, which had become popular a century before. Malthus had asserted that population tends to increase faster than the food supply. In the same two decades birth control advocacy took on additional arguments, as pleas were heard for the right of individuals to decide the size of their own families

and the expectation was voiced that the poor would choose to have smaller families if they could have the opportunity to limit births. Radical reformers and women's rights proponents demanded that birth control information be made freely available to all people, despite existing law to the contrary. The United States, for example, had restrictive state legislation, and the federal Comstock Law of 1873 banned the distribution of contraceptive information by mail. Some reformers, most notably the nurse Margaret Sanger, defied the law and provided this information. Sanger, who coined the term "birth control" in 1914, founded the American Birth Control League in 1921 and established the country's first birth control clinic in 1916.

In England, Marie Stopes became a vocal advocate during these same years. The racial concerns of many early reformers, as well as the focus on the individual's right to choose, were reflected in Stopes's clinics, affiliated with the Society for Constructive Birth Control and Racial Progress (founded by Stopes in 1922). For these reformers in the 1920s, birth control seemed a way to help improve the human stock, since improvement could be promoted by discouraging reproduction of unfit groups, as well as by breeding the most fit individuals. Biological theory and social issues were thus tightly intertwined, as the phrase "racial progress" indicates.

In the 1920s Sanger, along with advocates from the medical community such as the American obstetrician-gynecologist Robert Latou Dickinson, began to work systematically with social reformers to enlist biologists and physicians in the cause of birth control. Both Sanger and Dickinson presumed scientists to have the laboratory skills and biological knowledge needed to improve the available birth control methods. However, many of the biologists they recruited were explicitly interested in the evolutionary effects of birth control; these scientists supported the eugenic argument that reproduction of the genetically unfit ought to be curtailed. Thus, birth control efforts were aimed selectively at the poor or socially disadvantaged. Social class was presumed to be biologically determined. Between 1910 and 1930 this genetic assumption gained credence and prestige with the rise of the science of genetics, through the developing theory of the gene and contemporary discussion of the gene's role in inheritance. This limited knowledge of genetics was applied immediately to human populations.

The first purely scientific meeting to discuss the biological issues surrounding population growth took place in Geneva in 1927. The World Population Conference was organized by birth control advocates (including Sanger) to provide an objective scientific forum that could evaluate the existing biological evidence. Many well-known biologists and social scientists attended. A number of them were also eugenicists—that is, they urged genetic improvement of the human race through selective breeding.

Despite their guise of scientific evaluation of human genetics, eugenic arguments at this time were strongly influenced by subjective factors; learned, cultural traits were often mistaken for genetically determined characteristics. The eugenic arguments hid strong prejudices based on race and class—arguments that eventually, during World War II, were used to rationalize genocide, especially of Jews. These horrific events ultimately fostered antipathy to overt eugenic arguments for birth control. After mid-century, population control advocates began to emphasize once again the economic and social, as opposed to strictly biological, advantages of the small family.

However, the research efforts initiated in the 1920s and 1930s under the eugenics mantle ultimately yielded fruit in terms of more effective laboratory-proven birth control methods. The oral contraceptive pill and the modern intrauterine device (IUD), both introduced in the 1960s, led to active campaigns throughout the world to reduce high birth rates, especially in developing countries. Government efforts were particularly vigorous in India and China, where birth control information was made available in public clinics and a small-sized family was promoted by the government—the purpose being to improve the national standard of living by limiting the number of people dependent on the country's resources. Although the contraceptive pills, creams, and IUD's used were all developed in laboratory research, the success of these birth control campaigns was, like that of the campaigns of preventive medicine, dependent much more on social attitudes than on the state of biological knowledge.

Reproduction is a complex process that can be interrupted at many different points with many different political and social consequences. Each society must choose its own framework for defining which birth control methods will be most acceptable and most effective. This reality became apparent in the latter decades of the century in Third World countries' frequent rejection of Western birth control methods and the implicit social assumptions underlying them.

Nonetheless, the widespread public interest in population issues in the latter half of the century opened up new realms for biological research. Governments encouraged the scientific exploration of certain reproductive problems and funded research in specific areas of reproductive biology. Medicine, biology, agriculture, chemistry, and industry all contributed to and benefited from the further development of the reproductive science that spawned modern birth control methods.

The considerable public and political interest expressed in population control measures in the latter part of the century—and, indeed, from the beginning of the century—performed an additional service for biological research. It made apparent that scientific assumptions may often derive from culturally defined values that are hidden by claims of objectivity.

318. Sanger. A practicing nurse, Margaret Sanger in 1916 opened the first American birth control clinic in the Brownsville area of Brooklyn, New York, although disseminating birth control information was illegal. The photograph was taken when Sanger (*left*), her sister Ethel Byrne (*right*), and their colleagues were tried. Sanger served thirty days in the workhouse for maintaining a "public nuisance."

319. New York, 1916. Here is the Brownsville area in which Sanger opened her clinic. She viewed birth control as a way of alleviating the misery she saw in depressed neighborhoods like this—misery stemming largely from the prevalent poverty and the frequent deaths associated with large families and abortions. Influenced by the views of many biologists and physicians, Sanger later embraced eugenic arguments, which placed more emphasis on racial progress and human evolution than on the individual's right to prevent unwanted pregnancies.

320. Menu, 1927 World Population Conference. Sanger worked with the American experimental biologist
Clarence C. Little and the financier-philanthropist
Clinton Chance to assemble in Geneva biological, so-
ciological, and statistical authorities interested in research on human population issues.

321. Some scientific members of the conference.
Among those in attendance besides Little (*third from
left*) were the Italian social scientist Corrado Gini (*sixth
from left, with cane*), the British biologist Julian Huxley
(*fifth from right*), and the American biologist Raymond
Pearl (*fourth from right*).

322. Eugenics: testing for a sense of elegance. The subject was asked to examine "critically with hand and eye" samples of ten different kinds of fur arrayed on the table. The samples were to be arranged in order of elegance—how they would seem if made into a woman's "best coat." The test's preparator—Harry Laughlin of the Eugenics Record Office at Cold Spring Harbor, New York—presumed that the sense of elegance was genetically determined. The test was demonstrated at the Third International Congress of Eugenics, held at the American Museum of Natural History in New York City in 1932.

323. Frontispiece, English Eugenics Society pamphlet *The Aims of Eugenics*. The seemingly noble ideal of the eugenics movement—enhancement of human evolution—masked class, racial, and religious prejudice. While encouraging reproduction of the best human stock, eugenics sought to limit births of the unfit. The principles of animal breeding were assumed to be applicable to human society. Despite attempts to ground these ideas in the newly developing theory of the gene, such notions as quality and fitness were culturally based. Eugenic theory thus proved to have many political consequences, which came to light in the 1930s and during World War II when Nazi Germany enacted race laws and established extermination camps.

324. The small family: a biological view. "The greater the care, the fewer the young" was the caption accompanying this cartoon. After World War II, social and economic arguments for birth control replaced the genetic assumptions of the preceding decades. Nonetheless, the social arguments still made extensive use of various kinds of biological evidence.

325. Family-planning movement in India. One of the most populous countries in the world, India sought to limit the size of families through publicity campaigns featuring posters like this (showing an ideal family with just two children) and by setting up family-planning centers that provided advice and contraceptives. In the face of growing demand for scientific information on birth control, birth control technology became an active field of biological research in the 1960s and 1970s.

326. Intrauterine device. Use of the IUD is explained to a mother visiting a rural clinic in India's Madhya Pradesh State in the mid-1960s. The oral contraceptive pill and the modern IUD were both introduced in the 1960s, amid growing concern over global population growth.

327. Vasectomy for birth control. For men the Indian government offered sterilization operations as a means of birth control. The picture dates from the early 1970s. The intensity of the government's birth control program—people were sometimes pressured to undergo sterilization—encountered not a little resistance among the population and reportedly was one factor in Prime Minister Indira Gandhi's defeat in the 1977 elections.

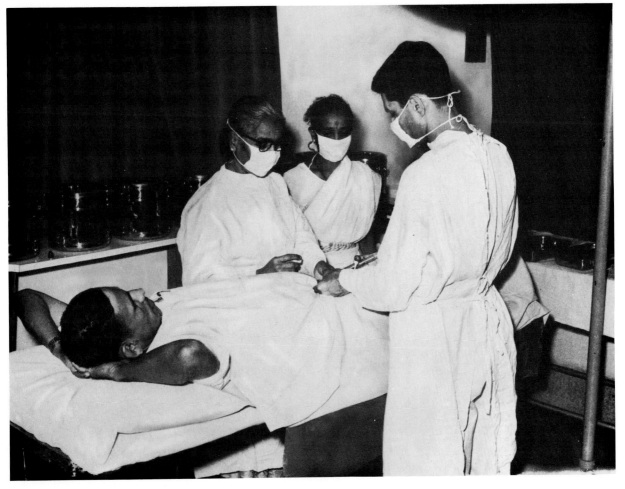

328. Sidewalk clinic in China. China, the most populous country in the world, began a series of birth control campaigns in the mid-1950s. The first two foundered on the rocks of domestic politics: the Great Leap Forward in the late 1950s and the Cultural Revolution in the latter part of the 1960s. Extraordinary success, however, was met by a wide-ranging program begun in 1971 that sought to limit families to two children. By 1979, China's birthrate had dropped to 18 per 1,000, from 34 per 1,000 in 1970. Because of the lingering severity of the population problem (aggravated by high birthrates during the Cultural Revolution), a program aiming at just one child per family, based on a mixture of rewards and penalties, was begun in 1980; after several years, however, it was relaxed. The picture shows doctors providing family-planning information at an informal outdoor clinic.

329. Family-planning exhibit at shopping center. Public education is a major priority in population control. This exhibit at Wuhan features a display of fetuses at various developmental stages. A doctor explains to schoolchildren the Chinese government's plan for limiting population growth.

330. Carrying the message to the public. The Chinese government made effective use of loudspeakers to promote its one-child-per-family program, particularly in rural areas, but also in cities like Zhengzhou, where this photograph of a loudspeaker truck was taken.

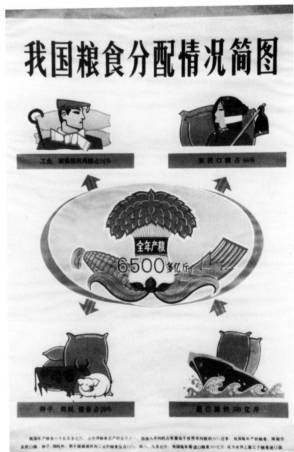

331. A comprehensive educational campaign. The Chinese family-planning program used virtually all forms of communication to convey a range of messages promoting the policy of limited population growth. Here, a poster tells how domestic and imported foodstuffs are distributed, the point being that lower population growth means a higher living standard for all.

332. Crop rotation research. Agriculture in the first half of the century was greatly influenced by research demonstrating increased production from the use of hybrids, fertilizers, and pesticides. However, systematic plant and stock breeding, as well as the introduction of synthetic chemical products, drew attention away from research on soil ecology and traditional modes of farming—such as this study of the effects of crop rotation at Lewiston, North Carolina. A reevaluation of agricultural practices followed from the ecological concerns that found expression in all areas of biology in the 1960s. By conserving soil, crop rotation can reduce dependence on fertilizers and pesticides.

23

Improving Agricultural Production

Agriculture was a significant contributor to, as well as a beneficiary of, the new experimentally based biology of the twentieth century. On the one hand, decades of systematic observations by farmers and herdsmen contributed greatly to the rise of genetics. Later, field studies of crop growth using various fertilizers, insecticides, and herbicides complemented laboratory studies of the biological effects of chemicals added to the soil or dusted on crops. On the other hand, agricultural productivity was greatly enhanced by scientific research—especially the laboratory work of geneticists, physiologists, and biochemists. Research on the nutritional content of livestock feeds, for example, radically altered agricultural practices and animals' production capacities.

The nineteenth century's evolutionary theory left to agriculture, as to human biology, a tantalizing legacy: the possibility of continued stock improvement. Breeders of crop plants and livestock were quick to realize this and early on allied themselves with the new genetic theories. Subsequent work at government agricultural research stations focused on how best to augment and enhance the improvement. Neo-Lamarckian and neo-Darwinian theories suggested different tactics: in the one case, environmental modification of the breeding stock; in the other, genetic modification. A kind of Lamarckian position prevailed for a while in the Soviet Union, under the leadership of the infamous Trofim D. Lysenko. He championed, for example, environmental modification of wheat by prolonged exposure to cold. Lysenko's views dominated Soviet agriculture until the 1960s.

Numerous studies conducted in the West, especially from the 1920s through the 1940s, indicated that specific varieties of commercial crops could be markedly improved by breeding. They could be bred, for instance, for high yield, or they might be bred for resistance to specific crop pests, which in turn affected yield. Hybridization of corn produced yields of reliable quality, providing an important economic incentive for further development of the science of plant genetics. Other plant hybrids proved commercially successful as well, resulting in increased sales of hybrid seed after World War II.

The century saw numerous synthetic chemicals developed specifically for agricultural applications. In the later decades, however, biological products were in some instances introduced in order to limit dependence on chemicals that had been found to be toxic. After World War II, for example, the use of newly developed insecticides and herbicides increased crop productivity. But pests could develop resistance to the chemicals. Moreover, the rising use of such chemicals and of chemical fertilizers posed serious environmental problems, especially through the accumulation of toxic substances in the food chain. The popular insecticide DDT, whose insect-killing properties were discovered in 1939, became an object of notoriety when its hazards were documented in the early 1960s by the American biologist and science writer Rachel

Carson in her best-selling *Silent Spring*. This enormously influential book led life scientists and others to realize that nonchemical methods, methods involving natural biological regulation, ought to be a better, safer means of insect control.

One nonchemical approach might be described as biophysical. It was found that reproduction in an insect species could be limited by releasing into the environment large numbers of the pests that had been sterilized by ionizing radiation. When radiation-sterilized males mated with females, no offspring were produced. Needless to say, this "autocidal" approach to insect control was usable only in cases where the target insects could be raised in substantial numbers, irradiated, and released inexpensively.

Another nonchemical strategy for insect control was the planting method known as intercropping. Alternating rows of a crop with rows of a different, naturally resistant species could prevent permanent infestation of fields without the hazardous side effects associated with chemical pesticides.

The regulation of animal growth and reproduction also became an important area of agricultural research during the century. In the first five decades agricultural production made notable advances thanks to the development of nutrition research, the discovery of vitamins and hormones, and the rise of reproductive technologies. Frozen bull semen began to be used for cattle breeding in the 1950s. In later decades experiments with superovulation and embryo transfer in cattle had significant commercial impact. These agricultural techniques were applied to human reproduction in the 1970s and 1980s.

The development of all these genetic, chemical, and reproductive technologies for agriculture was a consequence of twentieth-century optimism regarding the power of experimental science to improve the human condition. The innovations carried in their wake, however, serious and unexpected side effects. New problems, environmental and ethical, arose. The scale of the biological research enterprise, built on economic incentives in both agriculture and medicine, ensured that these unexpected results would have rapid and widespread repercussions in other domains. Dealing with this situation required rethinking the goals of biological research.

Nevertheless, the ecological interrelationships in the natural world remained a little-understood and largely unmapped territory within the life sciences. This continued to be so even though the sudden increase in human population resulting from the control of major infectious diseases put additional demands on agricultural productivity and sharpened concern for the balance of nature. Genetic manipulation, the product of the molecular biology revolution, provided but a limited solution to the complex environmental problems caused by human intervention in biological systems. The emphasis on controlling, rather than understanding, biological processes had created serious environmental consequences, whose solutions were not immediately apparent within the economic structures that supported the food production industries in developed nations.

333. Lysenko. The biologist and agronomist Trofim Denisovich Lysenko achieved virtually dictatorial control over agricultural and general biological research in the Soviet Union during the Stalin years and retained a position of power until the 1960s. Asserting a Lamarckian position in genetics, he rejected the chromosome theory of heredity; from 1948 to 1964, Mendelian genetics was not taught in the Soviet Union. Earlier, Lysenko had led an attack on the distinguished Soviet plant geneticist Nikolai Ivanovich Vavilov, who was arrested and exiled to Siberia in 1940. Lysenko had first gained the favor of Stalin by offering a faster way to improve crop production than the slow selective breeding and test trials advocated by Vavilov and other scientists. Lysenko, for example, sped the germination of winter wheat through vernalization, a cold treatment to break the seed's rest period. The effectiveness of such methods, however, was less than claimed. Lysenko rejected the value of inbreeding in producing hybrids. Because of his opposition, collective farms were for years not permitted to grow hybrid corn.

334. Breeding cotton for insect resistance. In the early 1920s a severe outbreak of leafhoppers (the jassid *Empoasca fascialis*) seriously damaged the South African cotton crop. At bottom is a plot of a commercial cotton variety that was severely injured by the insects. A nearby plot *(upper picture)* had a cotton variety called U4, specially selected for its resistance to jassids; despite the presence of the insects, the U4 grew normally. Later selections and crosses of this variety were made to enhance cotton quality and to achieve even greater levels of resistance. The two photographs date from 1926-1927.

335. Insect resistance in corn. Among the most destructive of plant-feeding insects is the corn earworm. An intensive search for varieties of corn resistant to the earworm included hybrid varieties. A 1939 experiment at Manhattan, Kansas, compared a moderately resistant inbred variety called Kansas K4 *(top)* with a highly susceptible inbred called U.S. 187-2 *(bottom)*. While moderate damage can be seen in the Kansas K4 ears, some of the U.S. 187-2 ears were completely eaten by the larvae.

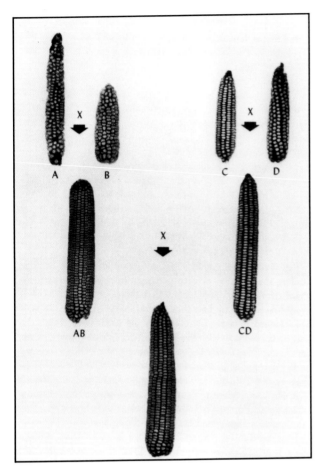

336. Line of descent of U.S. hybrid corns. After World War II, hybrid corn dominated maize production in the United States. Corn was the country's most important crop. Most of the hybrid corns used were the result of a double cross, based on research done early in the century by such scientists as George Harrison Shull and Donald F. Jones. The double cross uses four inbred lines—strains selected for desirable traits. (Crossing also restores vigor lost in the inbreeding process.) Thus, the double-cross hybrid ABCD was descended originally from the inbred lines A, B, C, and D. A and B were crossed, as were C and D. The single-cross hybrids were then crossed to produce ABCD, possessing the desirable traits of the original inbred strains. The hybrid corn's ears are better and more uniform in size than the ears of the original plants, and yields are larger.

337. Seed-corn drying plant. By mid-century this type of plant was used in the United States to process millions of bushels of hybrid-corn seed each autumn.

338. Herbicide research: 2,4-D. Using chemicals to control weeds is not a new idea, but progress accelerated after World War II with the introduction of so-called growth regulator herbicides, based on the chemical structure of plant auxins. A milestone was the entry into production in 1946 of the selectively acting growth regulator 2,4-D (2,4-dichlorophenoxyacetic acid), discovered in 1941 after a systematic search. In low concentrations 2,4-D was found to induce crippling abnormal growth, and eventually death, in most broad-leaved plants, while having little toxic effect on grasses. The photograph depicts the first experiment showing that 2,4-D could kill dandelions in a lawn without injuring grass.

339. First bioinsecticide. In the mid-1980s the U.S. Department of Agriculture announced the first bioinsecticide, effective against larvae of insects of the genus *Heliothis*, such as the corn earworm. At right, the water-soluble capsule held by the scientist contains enough of the insecticide to protect an acre of cotton against *Heliothis* caterpillars. Below, such a caterpillar, overcome by the insecticide, hangs from a cotton boll.

340. Biophysical insect control. Beginning in the 1950s some insect pests were successfully controlled by sterilizing males with radiation. Thus the screwworm fly, a livestock pest, was virtually eliminated from the southeastern United States by 1960. The photograph shows sterilized pupae of the tropical fruit fly being placed in a container for shipment to a test site in the Mariana Islands.

341. Cultural insect control: intercropping. Cultural control, using special methods of planting, growing, or harvesting, reemerged as an active area of research later in the century, when it was recognized that intercropping could reduce dependence on chemical pesticides. Visible in the photograph are rows of soybeans growing between rows of wheat stubble.

342. Irrigation. In this area of the Sahara Desert, formerly barren land has come to support crops as a result of an extraordinarily large center-pivot irrigation project.

343. Cow diet and milk production. Here is a moment in a mid-century study of how much of a cow's diet goes into milk production. Dairy cattle can be selectively bred for high milk volume and for milk rich in butterfat. These characteristics, however, are determined by many genes and are difficult to control. Environmental factors also influence the quality and quantity of milk produced. Thus, reliable metabolic studies are important to the dairy industry.

344. Computer monitoring in animal studies. In agricultural science, as in other life sciences, the advent of computers substantially enhanced researchers' ability to monitor their experimental subjects. Here, in a 1980s scene from the New York State College of Agriculture and Life Sciences at Cornell University, the computer tracks the cow's physiological and behavioral reactions to milking.

345. Upping cattle production: super-ovulation and embryo transfer. In the 1970s the U.S. cattle industry saw superovulation (inducing a cow to produce large numbers of eggs at one time through the administration of hormones) and embryo transfer become routine, producing thousands of calves a year. Above, the cow at right was superovulated and inseminated. Embryos taken from its uterus were transferred to the ten cows at left, which bore the ten calves.

346. Calves "manufactured" from a split embryo. By the early 1980s researchers reported success with this method of increasing livestock production. The twin calves Chris and Becky, born in May 1982, came from the same embryo and were thus genetically identical. Also shown is Colorado State University researcher Timothy J. Williams, who split the embryo when it was six or seven days old; the two halves were then implanted in the uteri of other cows.

347. The robot Maria. The idea of creating automata or artificial living
beings is an old one, but it enjoyed particular currency in the twentieth
century as the industrialization of society and the advance of technology
created the conditions for the robot to appear: first as a literary metaphor,
a symbol of certain developments in a changing world, and then as an ac-
tuality, of greater or lesser sophistication, in industry. The word "robot"
was first used, and applied to mechanical people, in the 1920 play *R. U. R.*
(Rossum's Universal Robots) by the Czech writer Karel Čapek; Čapek derived the word from the Czech *robota*, meaning forced labor. In the 1926
film *Metropolis* by the German director Fritz Lang, a mad scientist makes a
robot called Maria that endeavors to punish and destroy humankind. For
Maria, as for the monster made by Frankenstein in Mary Wollstonecraft
Shelley's 1818 novel, electricity was used to create life.

CHAPTER

24

*R*epairing
the *H*uman *B*ody

The century's scientific progress—science's growing mastery over nature and its strange and unexpected powers—fascinated and perplexed professional scientist and lay observer alike. Writers, especially of science fiction, explored possible ramifications of the powers that might be harnessed. One major theme was the creation of artificial beings—robots and bionic men and women. The dream of a mechanical body goes at least as far back as Homer, but it was in the twentieth century that the literature of the robot flowered. The robot or bionic human was a ready metaphor for a specter facing society: individuals endowed by science with physical and intellectual abilities that outstripped their emotional and ethical capabilities. The robot sometimes symbolized more as well—the potentially sweeping, and troubling, social impact of the century's explosion in science and technology.

But literature's artificial beings also pointed the way to the solution of physical problems encountered by humans in living out their lives. Medical science gradually entered this imagined domain by fabricating artificial body parts and using them (as well as body parts from human donors) to replace dysfunctional organs. By late in the century, readers were no longer surprised by artificial lungs, kidneys, and hearts. These had become realities. Science and technology had provided machines that could substitute for living organs.

The replacement of body functions by machines was envisioned early in the century. Indeed, machine imagery provided much of the analytical structure for thinking about how organs work. The kidney's role in purifying blood was studied by devising artificial kidneys that permitted waste products to be diffused through semipermeable membranes. After World War II, dialysis machines based on these same principles allowed people with kidney disease to survive without continual hospitalization. In the 1950s related devices were introduced for use in heart surgery to cool and oxygenate the blood while it was temporarily diverted outside the body during delicate surgical procedures.

One of the earliest substitutes for a living organ was the iron lung, developed in 1927 for artificial respiration of people with paralyzed respiratory muscles (such as polio victims). But it was a cumbersome external aid to the damaged body. The patient remained bedridden, and the machine dominated his or her daily experience, raising quality-of-life issues that would loom large in later decades as machines began to permeate medical practice.

The 1970s brought the first computer-controlled external devices that could replace muscles' motive power, compensating for the effects of war injuries, diseases such as polio, and the gradual physical decline accompanying old age. By substituting mechanical and electrical devices for the natural stimuli and motive power necessary to movement, they made it possible for, say, a paralytic to actually move about and climb stairs.

Machines were devised to perform other physiological functions as well. Pacemakers for the heart emerged as an important new technology around mid-century. The pacemaker generated an electric signal that was sent to the heart to ensure that it beat at a regular rate, thereby prolonging the life of an individual with an irreparably damaged heart. The first devices consisted of bulky electrical equipment. These were replaced by portable pulse generators, themselves later superseded by implantable electronic devices.

In legend the transplantation of body parts originated perhaps with Cosmas and Damian, the patron saints of surgeons. The two were said to have put the leg of a black man on a priest. But it was not until the twentieth century that surgeons were actually able to replace impaired organs with good, donor organs. Success with transplant surgery came slowly. It required development of specialized suture techniques, understanding of rejection phenomena that occurred during grafting experiments, and the capability to suppress the immune system's response to a graft. The cornea of the eye can be readily transplanted, with little fear of rejection, and indeed the first cornea transplant was performed early in the century. The transplantation of major organs like the kidney, liver, and heart did have to reckon with the immune response. The first successful kidney transplant came in 1954, the first liver and lung transplants in 1963, and the first human-to-human heart transplant in 1967.

The 1954 kidney operation avoided rejection problems because the donor was the patient's genetically identical twin. Transplants ceased to be a rarity only later, after the development of complex tissue-matching procedures and immunosuppressive drugs that could ensure the grafted organ would not be rejected by the recipient's body. The immunosuppressive cyclosporine was a major breakthrough, being less hazardous than earlier immunosuppressives; its general introduction in the 1980s opened up a new era in transplants.

Donor organs were less controversial than wholly artificial organs. The issue came to a head in the 1980s over experiments with "permanent" implantable artificial hearts. The first human to receive a permanent artificial heart was Barney Clark, a retired American dentist. After the mechanical heart was implanted in late 1982, Clark survived for nearly four months. The brief time lived by Clark and other recipients of permanent hearts was marked by limited mobility and by physical deterioration, raising ethical questions about the appropriate use of this technology. Artificial hearts were thus usually viewed as only "bridges"—temporary substitutes—until a suitable donor organ could be located. Nonetheless, they were the most technically sophisticated realization yet of the ancient dream of a mechanical body.

The robots, mechanical body parts, and other machines replaced elements of the natural order with artificial entities that blurred commonsense boundaries between living and nonliving things. These devices, such as respirators for comatose patients, gave rise to new ethical questions about the definition and rights of human beings—such as the right to die. Experimental science's probings caused these questions to be posed with increasing frequency as the century drew to a close.

348. Robot at an exhibition. This report from a 1928 issue of the *Illustrated London News* gives an idea of how the image of the robot had captured the public imagination. Early on the robot was used for publicity purposes. In this case, an aluminum robot—said to be capable of rising, bowing, and speaking—opened the Model Engineering Exhibition at the Royal Horticultural Hall.

A ROBOT TO OPEN AN EXHIBITION: THE NEW MECHANICAL MAN.

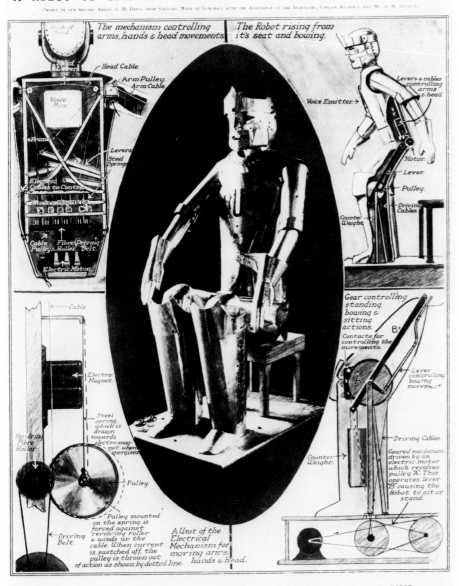

DRAWN BY OUR SPECIAL ARTIST, G. H. DAVIS, FROM SKETCHES MADE AT GOMSHALL WITH THE ASSISTANCE OF THE INVENTORS, CAPTAIN RICHARDS AND MR. A. H. REFFELL.

AN ALUMINIUM "MAN" THAT RISES, BOWS, AND MAKES A "SPEECH": A KNIGHT-LIKE ROBOT.

There has just been completed at Gomshall, near Dorking, the first British Robot, a gleaming thing of aluminium, not unlike a mediæval knight in armour, whose first duty will be to open a Model Engineering Exhibition to-day at the Royal Horticultural Hall. Concealed in the body is an electric motor which drives a fibre roller. Just above are several electro-magnets, with steel springs. To the base of these springs are fixed pulleys carrying cables that operate levers which move the Robot's arms and head. When the electro-magnets are energised the springs are drawn to the magnets pulling the edges of the pulleys against the revolving fibre roller. The pulleys revolve, winding in the cable and moving the head or limbs as desired. By cutting off current the wheel-face is detached from the roller, and the arm falls back to its normal position. For raising the Robot from its seat, causing it to bow to the audience and resume its seat, another motor is concealed in the platform below the figure's feet. This operates large pulley wheels concealed in the knees. When these wheels are slowly turned, a lever attached to each raises or lowers the man as required. Three contacts on the body and gives the "bowing" movement. To ease the work of the motor, counter-weights in the legs balance the weight of the body and interior mechanism. An ingenious electrical gear (which is the jealously guarded secret of the inventors) enables the Robot to hear questions and answer in a human voice. The Robot has been designed and made in under six months, so that it is but an infant and not yet able to walk, but the inventors state that in time it will be able to use its legs. At present, however, its chief work will be in the realms of publicity.

349. Replacing body parts. The machine functioning as an individual creature in its own right was not the only possible line of development that writers imagined for robotics. Also of interest, as this 1954 *Galaxy* cover suggests, was the replacement of worn-out, damaged, or otherwise unusable body parts with mechanical, electrical, or electronic equivalents. Prostheses are an old tradition: there is evidence that primitive artificial limbs were used in ancient times. But the period after World War II saw a sharp acceleration of development of replacement body parts.

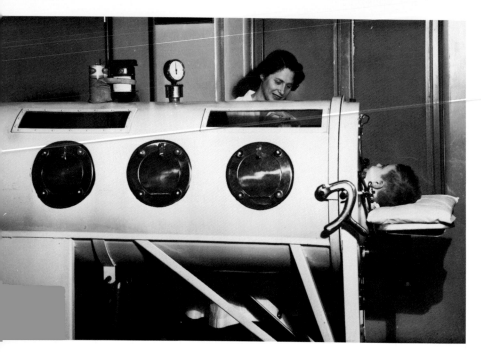

350. Iron lung. Developed in 1927 by the American industrial-hygiene specialist Philip Drinker, the iron lung was widely used for cases of paralyzed respiratory muscles—most notably polio victims. The machine applied alternately positive and negative pressure to the body's surface, forcing air into and out of the lungs. Although lifesaving in many cases, this cumbersome body-part replacement had disadvantages—the patient had to be kept under observation lest chewed food or vomit be inhaled. Also, prolonged confinement to a prone position could lead to the development of bed sores or infections.

351. Exoskeletal walking device. With the aid of this computer-controlled, hydraulically powered device, a paraplegic—a person suffering from paralysis of the lower half of the body—could stand up, sit down, walk, and climb stairs. The model shown here was developed by Ali Seireg and Jack G. Grundman in the United States. It was one of many mechanical walking machines tested in the 1970s and later.

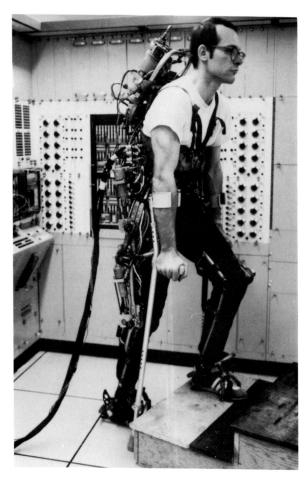

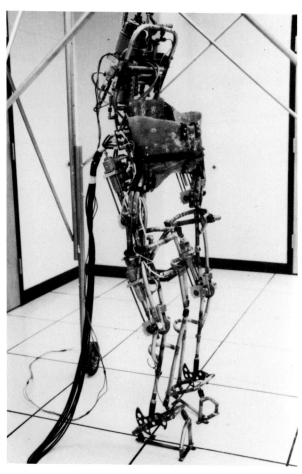

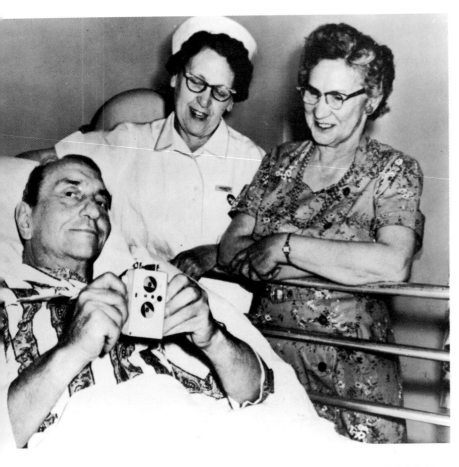

352. External pacemaker. Artificial electrical stimulation was first used to correct an irregular heartbeat in the 1930s, but for years the technique required bulky equipment carried outside the body. The patient shown here is holding a battery-powered pulse generator, developed by the American electrical engineer Earl Bakken in 1958. The generator could be attached to a belt worn by the patient. The electrode wire was passed through the chest to the heart.

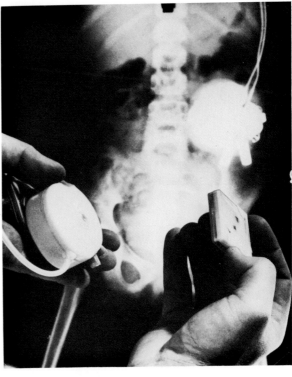

353. Implantable pacemaker. By about 1958, advances in electronic miniaturization had made possible the first long-term implantation inside the body of a pacemaker (including the battery power source). Refinements followed. The larger of the two implantable models shown here, the one on the left, can be seen in position inside the chest in the X ray in the background. The smaller pacemaker is the size of a matchbook. Such improved models were based on experimental studies in dogs and experience from clinical trials. In the 1960s the fixed-rate pacing of early pacemakers was replaced by variable pacing—that is, the rate at which the pacemaker delivered electrical impulses varied in response to the patient's activity level and physiological needs.

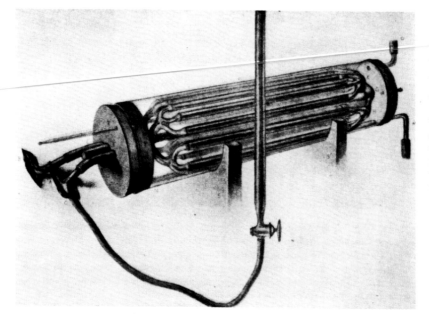

354. First artificial kidney. In 1912 the American biochemist and pharmacologist John Jacob Abel suggested using an artificial kidney to remove certain substances from the blood for study. The apparatus shown here was used with animals. For Abel this "vividiffusion" device was a research tool: with it he was able to show that free amino acids exist in the blood. But he did think it might prove useful as a clinical device, aiding people suffering from renal failure.

355. Inside Abel's vividiffusion device. The diagram at right shows how the system of tubes branched. The tubes were made of collodion—a form of cellulose nitrate—and were surrounded by a saline solution. Blood from an artery was fed into the device. As it passed through the tubes, small molecules filtered out into the surrounding fluid, while larger particles, like blood cells, did not. In this way—actually a form of dialysis—waste materials could be eliminated from the blood. The blood emerging from the apparatus was returned to a vein.

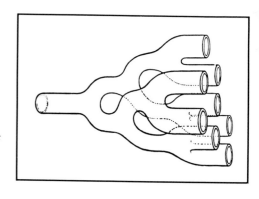

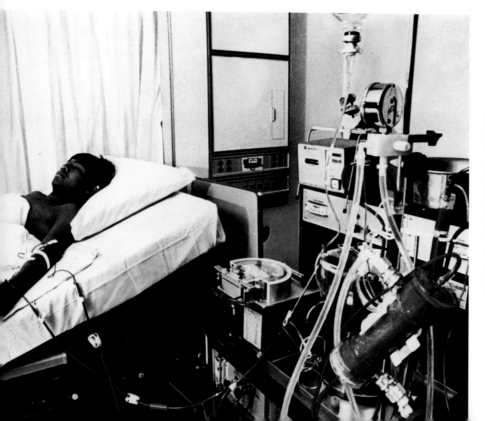

356. Artificial kidney in use. Abel's apparatus eventually led to the development after World War II of machines like this, capable of use on a regular basis for cleaning the blood of a person suffering from kidney failure. With the particular type shown here, used in the 1970s, the toxic wastes in the patient's blood were removed in four or five hours, the blood passing through the machine twice an hour. The procedure needed to be repeated three times a week.

357. Suturing and transplanting vessels. The French-American surgeon and biologist Alexis Carrel developed ingenious methods for joining blood vessels, methods he applied to the transplantation of vessels and, ultimately, entire organs. The three sketches illustrate (**A**) his transplantation in 1906 of a vein to connect the two ends of a severed artery; (**B**) the use in 1912 of a gold-plated aluminum tube for an artery "graft" (plastics came into use only later); and (**C**) the grafting in 1906 of a vein onto an artery—a technique subsequently used in coronary bypass operations.

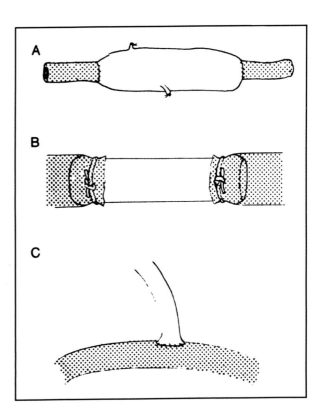

358. Baby Fae and the heart of a baboon. By the 1960s doctors had begun to transplant hearts. But a human donor was not always available. In 1984, Leonard Bailey, a surgeon at a California medical center, implanted a baboon heart in Baby Fae, born with a fatal heart defect. The implant was attacked by the immune system, and Baby Fae died in twenty days. Critics of the operation said that too little was known about the medical problems in interspecies transplantation and that rejection of the transplant was only to be expected. The few previous transplants of an animal heart to a human, the first in 1964, had involved adults and were failures.

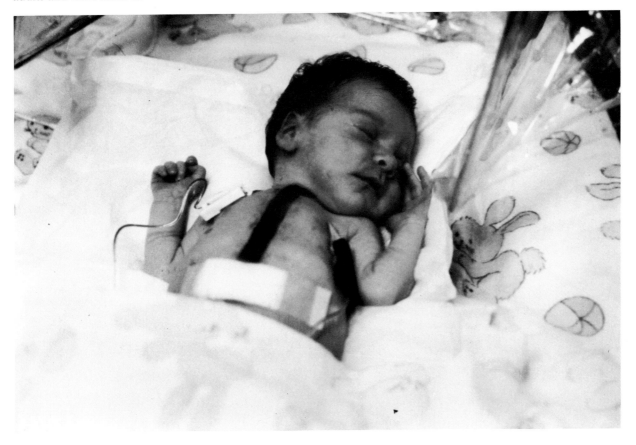

359. Kolff's artificial heart. Research on developing an artificial heart took two directions: one toward a device to assist the heart's operation and one toward a total replacement for the heart. The total replacement heart shown here, used in animal experiments, was developed by the pioneering program in the United States headed by the Dutch-born physician Willem Kolff. During World War II, Kolff had invented the artificial kidney for clinical use.

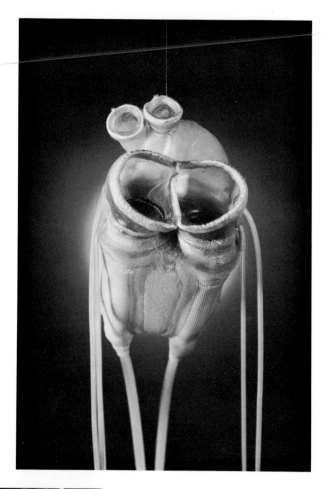

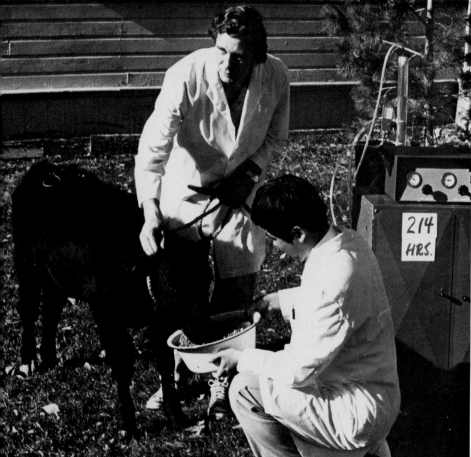

360. An early user. The calf in this early 1970s picture from Kolff's program at the University of Utah has been sustained by an artificial heart for 214 hours.

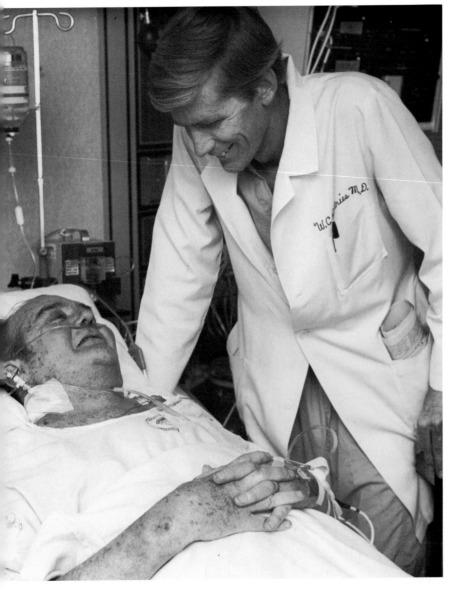

361. First human recipient of a permanent artificial heart implant. The first temporary implantation of a mechanical heart, for use as a "bridge" until a human heart became available, was by the surgeon Denton Cooley of Houston in 1969. A little over a decade later, in 1982, Barney Clark received a permanent artificial heart developed by Robert Jarvik of Kolff's team at the University of Utah. Here we see Clark and his heart surgeon, William DeVries, after the seven and a half hour operation. Clark died after nearly four months on the Jarvik heart.

362. Artificial womb. This apparatus, developed by the American physician Robert Goodlin at the Stanford University Medical Center in the 1960s, featured a stainless-steel tank, in which eight-to-ten-week-old fetuses resulting from abortion or miscarriage could be kept alive for up to two days. A deficiency of the artificial womb at this level of development was that it could not dispose of fetal waste materials.

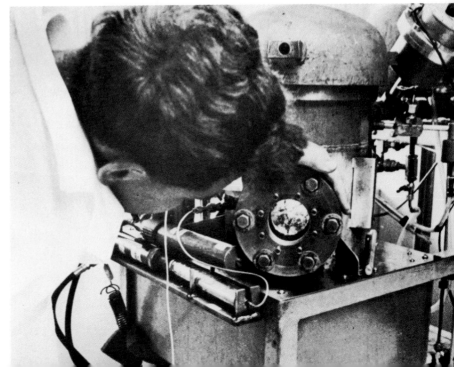

Part Seven

THE NATURE
OF THE
ORGANISM

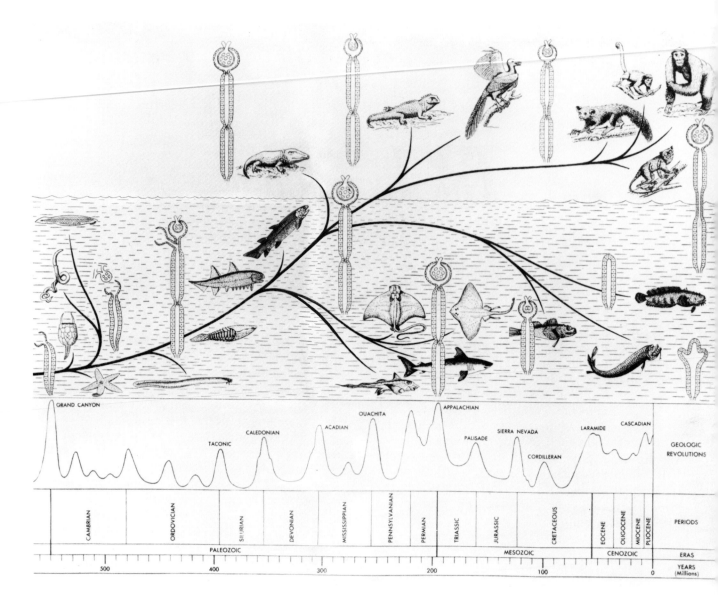

GRAND CANYON

OUACHITA

APPALACHIAN

CALEDONIAN

ACADIAN

SIERRA NEVADA

LARAMIDE

CASCADIAN

TACONIC

PALISADE

GEOLOGIC
REVOLUTIONS

CORDILLERAN

| CAMBRIAN | ORDOVICIAN | SILURIAN | DEVONIAN | MISSISSIPPIAN | PENNSYLVANIAN | PERMIAN | TRIASSIC | JURASSIC | CRETACEOUS | EOCENE | OLIGOCENE | MIOCENE | PLIOCENE | PERIODS |

| PALEOZOIC | MESOZOIC | CENOZOIC | ERAS |

| 500 | 400 | 300 | 200 | 100 | 0 | YEARS (Millions) |

363. The nephron through history. This mid-century diagram plots the evolution of the nephron—the basic unit of the kidney—against a background of the unfolding of geologic time. The nephron is depicted in a very schematic way, with straightened tubule. Early nephrons had just a tubule. Later nephrons of land animals had an intricate glomerulus, or tufts of capillaries, and a two-segment tubule. The animals are shown in the order of their development; a saltwater environment is at the far left and at the right, and a freshwater environment, indicated by less closely spaced lines, is in between. The emphasis here is on change of form and function throughout the evolutionary history of life. The data are presented in biologists' traditional format—that is, the changing organization and structure of organisms over time.

25

The Language of Biology

The everyday language of biology changed markedly during the century. The visual data of observation and description largely gave way to numerical and graphic data based on the methods and techniques of the physical sciences. This transformation was due not only to the rise of the experimental method and the introduction of new instruments and analytical procedures but also to the changing focus of biological research. In particular, biologists in the second half of the century turned more and more from study of the organism to study of its constituent molecules. Consequently, geometric representations of molecular structure, derived from physics and chemistry, increasingly informed biological discussions.

Traditionally, biologists examined the form of organisms and the relation of that form to function. Study of plants and animals in morphology, anatomy, embryology, histology, and cytology yielded descriptions of plants, animals, microorganisms, organs, tissues, and cells and comparisons of structurally similar organisms. Biologists speculated about the relations between organisms and correlated their interpretations with historical change in the earth's surface, thereby refining the concept of biological species. The theory of evolution had strengthened the focus on change in form over time. Rather than merely enumerating and describing living organisms and their parts, twentieth-century biologists emphasized organisms' interrelatedness and documented specific transformations over time.

Early in the century, studies of function stressed the importance of this temporal sense. The graphing of change over time increasingly entered the routine vocabulary of biology. Data that had once, in the nineteenth century, been presented only as tables of numbers now came to appear regularly as graphs. Electronic descendants of the revolving-drum recorder expressed dynamic change in wavy lines that could be analyzed by the tools of mathematics. This mode of data acquisition and presentation was exploited especially by physiologists and biochemists, who were bringing biology into closer ties with the methods and goals of the physical sciences.

The powerful new analytical techniques, however, did not eliminate biologists' preoccupation with structure, be it of organisms or of molecules. Rather, biological concepts like the unit of heredity, the gene, were translated into chemical terms and into the vivid geometric language of the chemist. Analysis of structural and functional components of organisms followed the chemists' familiar pattern of isolation, purification, and crystallization of discrete molecules, whose structure could then be determined. The relationship between chemical structure and physiological function was in turn examined further in simplified experimental systems.

During the century physiologists, chemists, and physicists studying different aspects of life processes analyzed the chemistry of the complex organic molecules called proteins. Correlating biological, chemical, and physical data led to the construction of geometric models of molecules associated with major biological phenomena like heredity, reproduction, respiration, and growth.

Researchers began understanding biological processes essentially as complex sequences of chemical events, in which different atoms interacted to produce recognizable change at another, higher level of organization—the molecule, the subcellular particle, the cell, the tissue, or the organ. This reduction of complicated biological events to chemistry and physics had long been the goal of experiment in biology. Late-twentieth-century biology, especially, reflected the hope that life processes would ultimately yield their secrets to the analytical techniques of the physical sciences. Nonetheless, the process of discovery in biology was by no means simply a matter of cookbook analysis using the techniques of physics and chemistry. Correlating data from widely disparate areas of research often shed unexpected light on complex biological phenomena—as in the discovery of the double helix of DNA.

Evolutionary theory provided a unifying conceptual thread in biological research. But by mid-century this approach, too, became focused on molecular structure. Scientists' belief in the laws of nature and the unity of observed phenomena led them to take organisms apart in order to understand how they work. In the wake of the exhilarating discoveries and predictions of molecular biology, study of the whole organism and the functioning ecosystem was usually subordinated to this molecule-oriented agenda.

Chemistry had become the common language unifying the physical and biological sciences, the code in which the properties of life were expressed. It was a system of symbols whose use translated the very complex processes of living organisms into a language accessible to biological and physical scientist alike. Transient, internal events were represented in the interactions and transformations of nonliving molecules whose joint properties represented the metabolic and sensory capacities of animated beings. The success enjoyed by this reductionist approach, however, diverted attention from the more complex and unique features of living organisms, such as consciousness, behavior, and social organization.

364. Henderson's nomogram. A convenient way of visualizing the relations between several variables—a situation arising in a complex, multicomponent system—is the nomogram, introduced into biology by the American biochemist Lawrence Joseph Henderson in his work on blood in the 1920s. In the diagram at right Henderson presents the seven variables he used to explain the respiratory activity of mammalian blood. Each point has seven coordinates; knowing two means that the other five can be read from the nomogram. The chart portrays the effect of a change in one variable on the others. In this chemical system the physiological properties of blood gases are affected by their changing pressure.

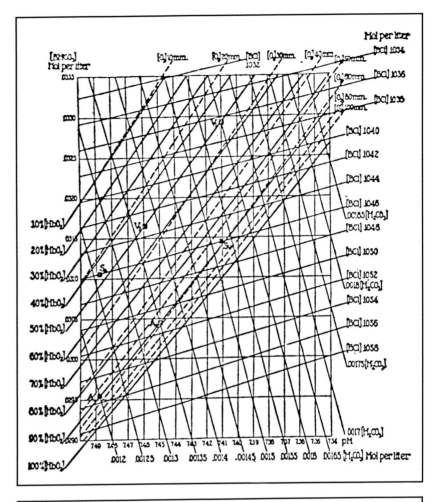

365. Human EEG patterns. Recording the human brain's electrical activity with electrodes on the scalp, a technique introduced in the late 1920s, exemplifies biologists' efforts to monitor physiological change over time. The electrical activity is symbolized graphically by waves representing changing bioelectric potential. In each of the pairs at right (from a 1970s text) the top tracing is from the left frontal region of the brain; the bottom, from the right occipital region. The first pair provides a basis for comparison—the subject was awake and alert. In the other cases the subject was (b) drowsy; (c) in light sleep, with occasional slow waves; (d) in slow-wave (deep) sleep; (e) in deeper slow-wave sleep; and (f) in rapid-eye-movement sleep, when most dreaming occurs. The form taken by the brain waves reflects the changing state of the whole organism and the presence of complex phenomena like sleep and dreams.

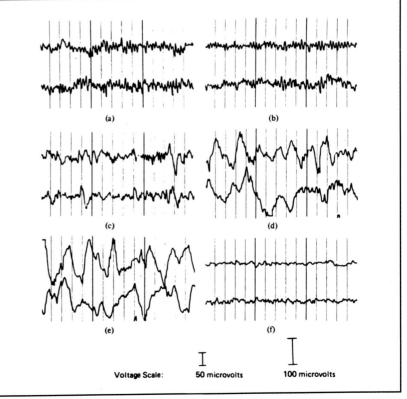

(a) (b) (c) (d) (e) (f)

Voltage Scale:　50 microvolts　100 microvolts

366. Crystallized enzyme: trypsinogen. Crystallization was a crucial step in the process by which many chemicals in living beings were isolated and characterized. In 1926 the enzyme urease, from the jack bean, was isolated, crystallized, and characterized as a protein. The 1930s saw the crystallization, and identification as proteins, of a few more enzymes, including the digestive enzymes trypsinogen and trypsin. Trypsinogen is actually an inactive, or precursor, enzyme; it is produced in the pancreas and converted into the active enzyme trypsin in the intestine.

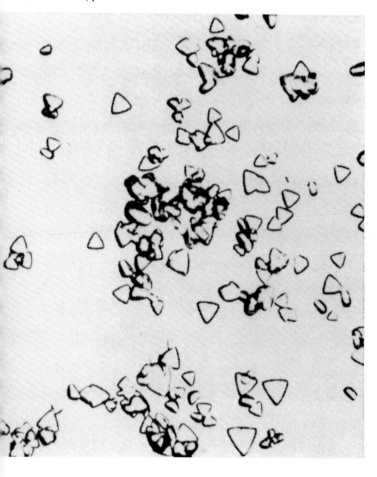

367. Crystallized enzyme: trypsin. Compare this with the preceding photograph. The remarkable differences in structure are due to the chemical changes wrought in activating trypsinogen. The middle decades of the century saw notable advances in elucidating the correlation between chemical structure and biological activity, as the molecular structure of proteins was gradually pinned down. Both photographs were made in 1935 after the two enzymes were obtained from cattle and crystallized.

368. Building a gene. In 1970 a team at the University of Wisconsin headed by the Indian-born biochemist H. Gobind Khorana made the first synthetic gene, a copy of a yeast gene. Here we see Khorana explaining the technique by which the gene's seventy-seven component nucleotides were chemically assembled. The synthesis was based on the analytical sequencing work of the American biochemist Robert W. Holley. Such mastery of the techniques of analysis and synthesis of the major biological molecules was a principal goal of twentieth-century biology and biochemistry. Holley and Khorana shared the 1968 Nobel Prize in physiology or medicine with another American explorer of the genetic code, Marshall W. Nirenberg.

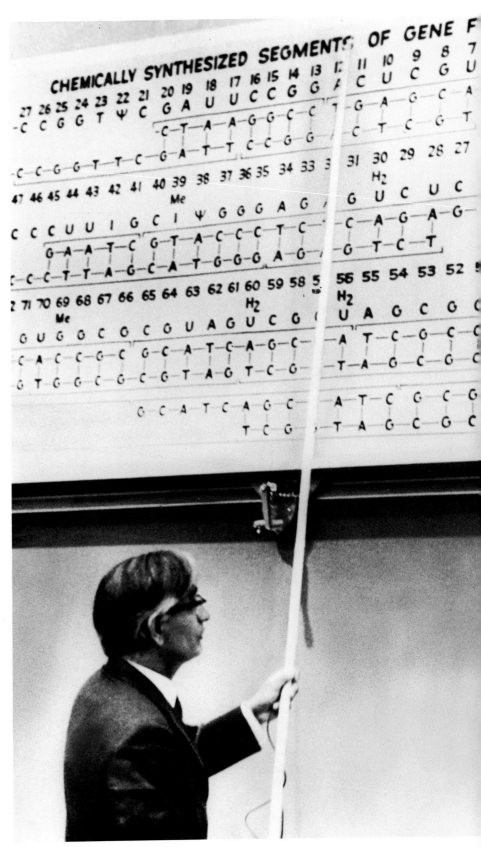

370. DNA double helix. Pauling was one of several researchers who attempted to decipher the structure of DNA. Despite his success in working out the alpha helix structure of proteins, he did not hit on the double helix form that underlay the structure of DNA—that insight was due to Watson and Crick. The structural model below is more schematic than Pauling's molecular representations. The chains of DNA can be readily seen; they wind about a fiber axis. From each "pentose" sugar (represented by a pentagon) on the chains, there extends a base, which connects with a base on the opposite chain by a hydrogen bond. It is these relatively weak hydrogen bonds that hold the chains together. The double helix structure allows replication of the genetic information encoded in the sequence of bases along each chain.

369. Pauling. The American chemist Linus Pauling made seminal contributions to knowledge of chemical bonds and their relations to the properties of matter, including organic matter. In 1948 he saw in a flash of insight that some protein chains of atoms are arranged in a helical manner—the so-called alpha helix. This early 1950s photograph juxtaposes Pauling with a knob model of a complex molecule having a helical structure; he was responsible for introducing these models to depict the three-dimensional structure of molecules. Pauling received the Nobel Prize in chemistry in 1954 for his work on chemical bonding and the Nobel Peace Prize in 1962 for his efforts toward the banning of nuclear weapons tests.

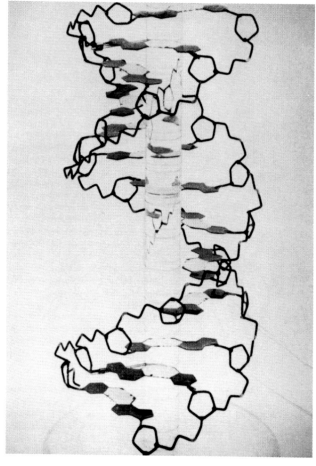

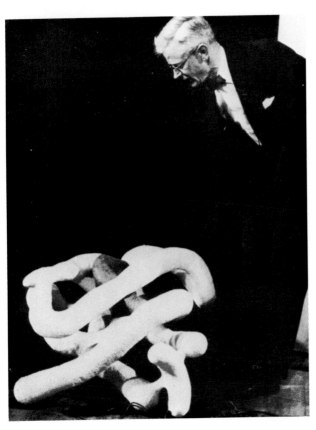

371. Kendrew and myoglobin. The British biochemist John Kendrew built this low-resolution plasticine model of sperm whale myoglobin—an iron-containing protein found in muscles—after extensive X-ray studies led him to the basic structure in 1957. It was the first time the three-dimensional structure of a protein had been solved. Kendrew was surprised to see how irregular it was. The gray disk half-hidden at the top (*center*) of the model is the heme group, a set of atoms centering around an iron atom. The model's long straight sections were found to be alpha helices.

372. Human hemoglobin. This was the first high-resolution model of the atomic structure of oxyhemoglobin. It was built by the Austrian-born British molecular biologist Max Perutz and his colleague Hilary Muirhead in 1967. The vertical rods were for support. The inset shows the crystalline structure of sodium chloride, or common salt, to the same scale—that is, 2 centimeters per angstrom. Note how much more complex hemoglobin is than salt. Hemoglobin, the protein that transports oxygen in the blood, has ten thousand atoms.

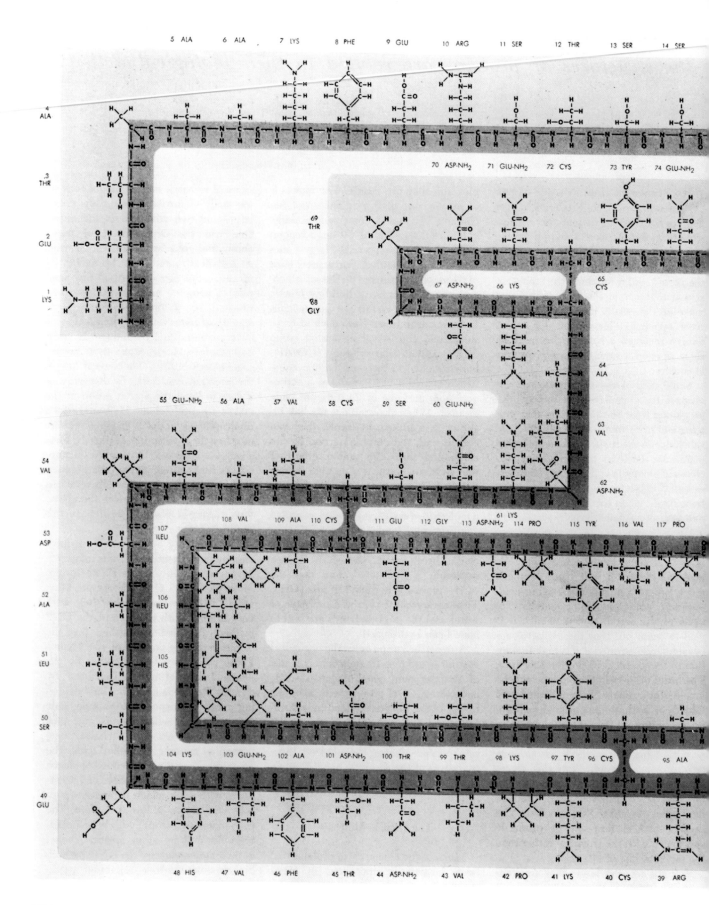

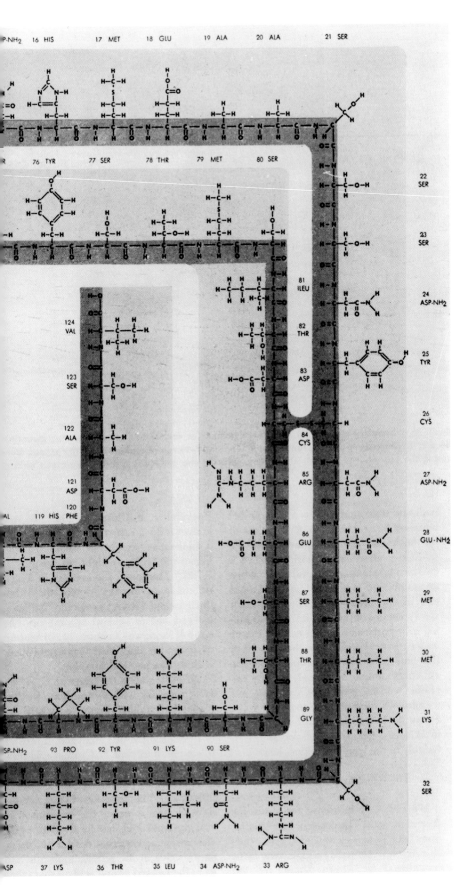

373. Ribonuclease. This two-dimensional diagram from around 1960 presents the structure of a molecule of the enzyme ribonuclease, which digests ribonucleic acid (RNA). Note that the molecule consists of a chain of amino acids. Stable cross bridges link parts of the chain into a compact (and ultimately three-dimensional) structure. The sequence of the 124 amino acids was determined by analysis. When artificially synthesized, this molecule possesses all the properties of the naturally occurring enzyme, reinforcing many biochemists' claims that life results from the properties of molecules. This was the first enzyme to be synthesized by the methods of organic chemistry. The amino acids appearing in the molecule are indicated as follows:

ALA	alanine
ARG	arginine
ASP	aspartic acid
ASP-NH$_2$	asparagine
CYS	cystine
GLU	glutamic acid
GLU-NH$_2$	glutamine
GLY	glycine
HIS	histidine
ILEU	isoleucine
LEU	leucine
LYS	lysine
MET	methionine
PHE	phenylalanine
PRO	proline
SER	serine
THR	threonine
TYR	tyrosine
VAL	valine

374. Checking the apparatus. Increased use of instruments and a concomitant concern with rigor and quantification have marked twentieth-century biology—a sharp contrast to the less precise and natural history–oriented researches of preceding centuries. This 1938 photograph catches the British biologist Julian Huxley (*foreground*) and Ludwig Koch in the recording van they used in a study of animal language.

26

The New Research Enterprise

The century saw standardized recording techniques and specialized analytical methods introduced into virtually all areas of biology, changing the practice of research and altering the everyday experience of biological scientists. Most often scientists now performed their labors in a uniquely defined workspace called the laboratory rather than out in the natural world, in the field. They regularly studied organisms in controlled, well-defined environments using a vast array of specialized equipment that could stimulate, monitor, or record the phenomena under investigation. Animals, plants, organs, tissues, cells, subcellular particles, and enzymes were put into boxes, containers, and test tubes where they were maintained, manipulated, and analyzed. All this made for an increasingly precise biology, but not necessarily a biology that would be responsive to human needs.

Culture techniques perfected in the second half of the century gave rise to new technologies for replacing defective organs and genes. These technologies, however, provoked concern among the public about the direction biological research was taking. In the first half century there had been a string of potent new chemicals, such as antibiotics, vaccines, fertilizers, insecticides, and herbicides. But now scientists envisioned developing artificial organisms that could produce desired molecules or utilize certain industrial by-products. And the new transgenic organisms made by gene-splicing techniques posed serious questions. Could they be a threat to public health? Who could be said to own them, and to be entitled to profits they might yield? New regulations were introduced for safe containment of the research work, as well as new laws defining the status of the products of biological research. Genetic engineering, like human reproductive technology, juxtaposed the public's ultimate hopes for science and its ultimate fears—that is, optimism that natural processes could be controlled for the benefit of humanity and apprehension that human understanding of a complex world was still so limited that a Pandora's box of unforeseen ills would be opened.

Biological science had become industrialized. Big science, industrialized science, promised bold new products in the late twentieth century: cures for disease, replacements for worn-out organs, technological fixes for complex social and economic problems like famine and global warfare. But the industrialization of the research environment undermined the tradition of disinterestedness that had been basic to Western science. It was in this problematic setting that scientists continued to explore the unknown.

375. Organ bank. The animal-organ bank above, located at the University of Minnesota, was built not long before the photograph was taken, in the mid-1960s. The apparatus stored organs in oxygen and under pressure at a temperature of about 39 degrees Fahrenheit, for eventual use in experimental studies of organ transplantation.

376. Pathogen-free animal room. This controlled environment at the Jackson Laboratory in Bar Harbor, Maine, houses 25,000 mice used for experimental studies of animals raised free from specific microorganisms. The temperature and humidity are regulated, as is the ventilation. A positive air pressure is maintained to prevent infection from being carried in from the outside.

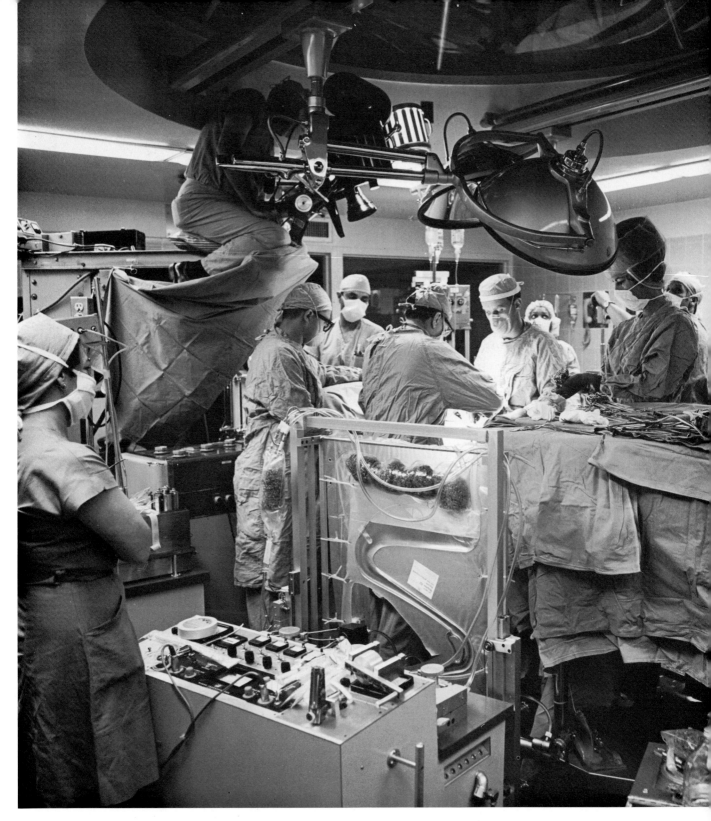

377. Open-heart surgery. The surgeon is Michael DeBakey, who pioneered
a number of techniques for correcting heart and circulatory problems.
Since the heart is stopped during such procedures, using these sophisticated surgical techniques requires the support of devices, like the heart-
lung machine, that maintain the circulation and oxygenate the patient's
blood.

378. Genetic engineering lab. With the advent of gene-splicing in the 1970s, the United States adopted a system of containment standards. The highest, P4, was equivalent to the containment level recommended for work with the most lethal microorganisms and viruses. The laboratory above, at the U.S. National Institutes of Health, met the P4 requirements for recombinant DNA research.

379. Industrial research. Large-scale genetic engineering and other biological research came to be done not only in government and university laboratories but also by private industry. The greenhouses shown below are atop the Monsanto Life Sciences Research Center near St. Louis, Missouri. The center was established in 1984 to create new drugs, crop plants, and microbial pesticides.

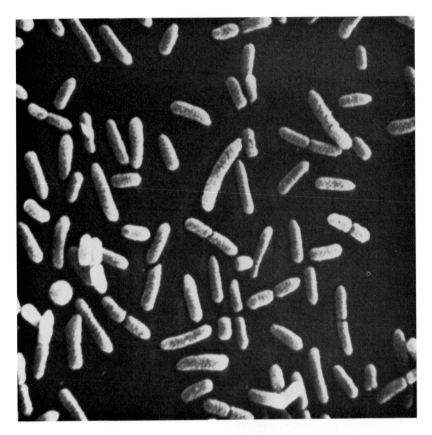

380. Oil-eating bacteria. Designed to combat oil spills by converting hydrocarbons into carbon dioxide and proteins, these bacteria were developed at General Electric in the early 1970s by Ananda Chakrabarty, a biochemist who combined genetic material from four different strains of the bacterium *Pseudomonas*, each of which was able to digest a different component of crude oil. The new strain was the cause of a U.S. Supreme Court ruling in 1980 that "a live, human-made microorganism is patentable subject matter."

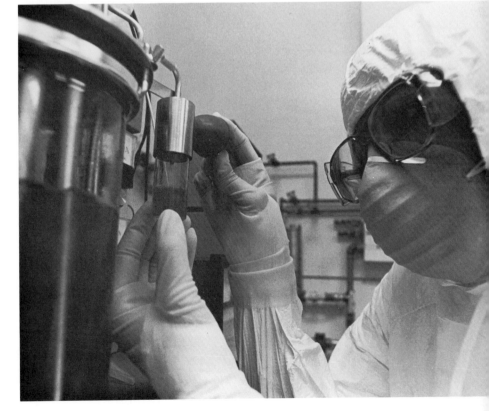

381. Genetically engineered vaccines. Here a scientist at the Merck Sharp & Dohme laboratories is drawing a sample from a small fermentation unit used in the development of the first American genetically engineered vaccine for humans. The vaccine, Recombivax HB, protecting against infection with the hepatitis B virus, was licensed by the U.S. government in 1986.

382. Minamata disease. The biological effects of industrial effluents caused increasing concern in the century's final decades. A sequence of events unfolded around Japan's Minamata Bay in the 1950s that exemplified the potentially disastrous consequences of the careless use of chemical technology. People began exhibiting such symptoms as sensory disturbances, delirium, paralysis, and brain damage; some died. A lengthy investigation revealed that a chemical factory producing acetaldehyde and vinyl chloride used a mercury catalyst. The plant released mercury wastes into the fishing waters of the bay, and fish became contaminated with methyl mercury. People who ate fish from the bay suffered mercury poisoning. Over 1,500 persons were affected. In this classic photograph by W. Eugene Smith, a girl congenitally maimed by mercury from the contaminated fish is bathed by her mother.

27

*O*ur *P*lace in *N*ature

Living organisms and ecosystems are extraordinarily complex. Individual researchers made this point emphatically and repeatedly throughout the century, but it was widely ignored because of the seductive simplicity and power of manipulative laboratory techniques, which generated new drugs, vaccines, fertilizers, and hybrids for human benefit. In the process, the biological sciences grew rapidly and expanded—but not without unforeseen social consequences. The optimism that so characterized the century's early years was gradually replaced by doubt, by fear that human tinkering with the forces of nature, still incompletely understood, might have damaging effects. In the latter part of the century some biologists and physicians, along with some physical scientists, voiced concern that scientific knowledge might be serving inappropriate ends. More and more people began discussing basic ethical questions raised by the direction that biological research was taking.

Much of this new concern grew directly out of the trauma of World War II. The Nazis employed biological arguments of racial superiority to justify mass exterminations. Concentration camp inmates were used as experimental subjects for physiological and surgical research. The revelations of these horrific experiments vividly demonstrated people's inability to foresee the full moral consequences of scientific research. So did the gradual identification, years after the war, of the biological and medical effects of the dropping of the atomic bomb. Radiation burns, radiation sickness, and the long-term genetic effects of nuclear radiation were the legacy and the promise of continued testing of nuclear weapons. Neither plants nor animals, including the human race, could escape an environment permanently altered by radiation from further weapons development.

In some parts of the globe the environment had been intentionally altered in order to bring water or civilization to uninhabited areas. But the movement of human activities into these pristine surroundings befouled the water and dirtied the air. Acid rain spoiled lakes and forests. Toxic runoffs threatened nearby communities. Chemical products that had been developed for useful agricultural purposes turned pernicious. Some, designed to eliminate pests and weeds and thereby raise yields, accumulated in the environment or the food chain and posed a continuing threat to public health. DDT was a potent insecticide, but it left in the environment a residue so hazardous to wildlife that many countries restricted or banned its use. The advent of gene-splicing techniques late in the century presented genetic and environmental hazards in a new guise, although experts argued that the potential benefits from such research would far outweigh the potential hazards.

Late in the century, biologists such as these experts, confident of their sophisticated technical skills, were not yet persuaded of the wisdom of pursuing some of the classical biological questions that had been swept aside by the excitement of laboratory studies. While it was becoming more and more imperative to take up once again the question of humanity's place in nature, biology had grown firmly rooted in the laboratory, rather than in the field. The desire to gain complete mastery over the environment and to cure disease had shifted many scientists' attention from understanding nature to controlling and manipulating it.

These scientists believed in a simplistic form of human progress. Their definition of progress was a direct legacy of the optimism in late nineteenth-century Western society regarding the potential fruits of laboratory science. In the twentieth century the steady outpouring of new biological products—vaccines, hormones, vitamins, antibiotics, synthetic drugs, insecticides, herbicides, contraceptives, hybrid seed, transgenic organisms—reinforced the long-voiced conviction that the way to economic and social progress lay in supporting and developing experimental research. Yet the scientific method, based on posing questions to nature, also carries within it the possibility of new directions and new emphases. Scientists have choices to make regarding what they do, what they study, and what they produce. In the biological sciences these options constantly confront researchers through the central dichotomies of reductionism and holism, quantity and quality, cure and prevention, and exploitation and conservation of natural resources.

Toward century's end, controlling the environment and curing all disease appeared increasingly to be unfulfillable goals. To many it also became clearer and clearer that continued efforts to reach these goals were fraught with unanticipated risks. Broad advances in curing infectious disease, for example, were followed by increases in chronic disease. New technologies for warding off death brought continued life to some, but at the price of continued suffering, or at least bondage to the life-giving machine. Progress against killer diseases in developing countries led to a population explosion, contributing in some cases to widespread famine. The emerging concerns over the unforeseen social and environmental hazards of new technology promised to shape key arenas of research for the twenty-first century.

Since the products of laboratory research were partly responsible for widespread pollution, unemployment, hunger, and popular disaffection with science, biologists could ill afford to neglect reflecting on the industrial agenda that had drawn them from the field to the laboratory. Research protocols, grant proposals, and national science policies all contained implicit social goals hidden within the objective appearance of research science. Biological scientists' own research and experience had laid bare many of these concealed assumptions. Through the mid-century revolution that had brought the biological sciences to prominence, biological scientists had gained the power, and with it a social obligation, to initiate and shape discussion of these concealed assumptions and to suggest alternative strategies for the future. For the twenty-first century, as for the nineteenth and twentieth, the central question would remain the place of *Homo sapiens* in nature.

383. Human beings as experimental animals. Some of the most notorious violations of traditional moral and ethical restrictions on experimentation with human subjects took place under the Nazi regime. This picture of a freezing experiment on a political prisoner was taken at the Dachau concentration camp, established in 1933. Observing the subject, who is immersed in ice water, are E. Holzloehner (*left*), professor of physiology at the medical school of the University of Kiel, and Sigmund Rascher.

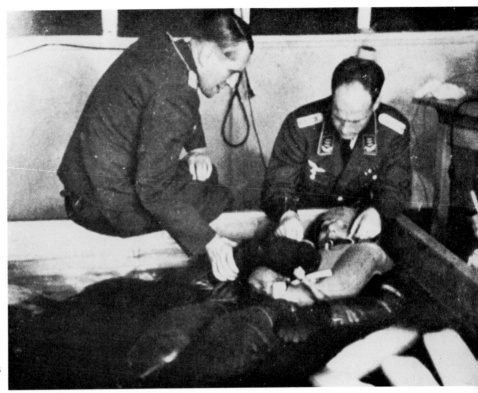

384. Nazi concentration camp operating theater. The camp is Auschwitz (Oświęcim) in Poland. Castration and sterilization experiments on prisoners were common, reflecting concern for racial purity as well as interest in the reproductive mechanism.

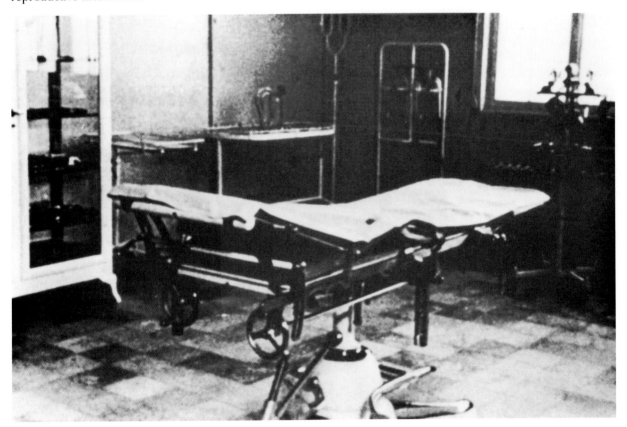

385. The advent of nuclear weapons. World War II saw the invention of the atomic bomb, which represented a marked escalation in human beings' capacity to destroy themselves, not to mention disrupting the entire biosphere. In this cartoon, according to the caption, one fly says to another: "There's one advantage in our inability to make decisions—we never make the wrong ones."

386. A-bomb victim. Not only was the atomic bomb invented during World War II, it was used, twice. In August 1945, U.S. planes dropped A-bombs on Hiroshima and Nagasaki in Japan. Both cities were virtually destroyed by the explosions and resulting fires; in each, tens of thousands of people were killed or injured. In addition to the major injuries due to shock waves, the collapse of buildings, and fire, many people suffered radiation poisoning, whose effects may be immediate or long-term. The photograph was taken in September 1945 in a makeshift hospital at a bank building in Hiroshima.

387. Altering the environment. Here are two views of a locality in Colorado: above, Clear Creek in 1873; below, the area in 1977, after a dam was built. The effects of human intervention in the environment may be intentional and overt, as here, or unexpected and multifaceted. Chemical residues accumulating from pesticides and industrial waste, as well as genetic changes induced by radiation, affect organisms throughout the food chain and permanently alter ecological relationships. While laboratory experiments may hold constant all variables but one (in order to study its effects), perturbation of the natural environment can have profound and unexpected effects on the entire ecosystem.

388. Fighting the effects of acid rain. The 1980s saw growing fears that an influx of acid from the atmosphere was disrupting delicate ecosystems such as those of lakes. Atmospheric acid stems partly from natural sources, like volcanic eruptions and forest fires, but largely from smokestack emissions of industry. Since hard-water lakes containing carbonates were less affected by acid rain, Swedish researchers sought to reduce the acidity from acid rain in other lakes by adding carbonate-containing lime to them. A possible danger, of course, was that the lime might itself disturb the lake's ecosystem. The Swedish lake above is receiving a ton of lime a minute.

389. Monitoring acid rain. Here, rainwater is being collected, and its acidity analyzed, near an Adirondack lake in New York State. The Adirondack Mountains were hard hit by acid precipitation.

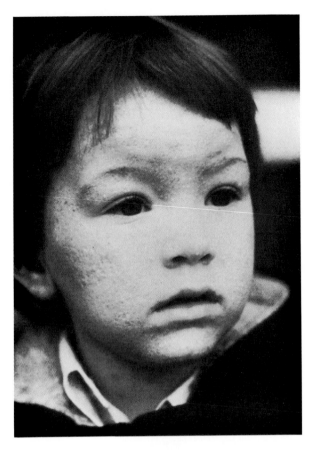

390. Dioxin victim. The toxic effects of a class of chemicals called dioxins drew public concern in the 1970s and 1980s. Produced as by-products in a number of industrial chemical processes, dioxins vary in toxicity, the most dangerous to humans being the dioxin known as TCDD. The girl in the photograph is suffering from the painful acnelike skin disease chloracne, caused by exposure to dioxins released in an explosion in 1976 at a chemical plant near Seveso, Italy. Containment of the products of new chemically based industries presented a formidable task for twentieth-century governments.

391. Landscape defoliated by Agent Orange. During the Vietnam war the United States used the dioxin-containing herbicide Agent Orange to remove crops and forest cover that could be of benefit to the enemy. But not only the landscape may have been laid waste by this tactic. Some veterans of the war claimed to have contracted cancer and other diseases from exposure to Agent Orange; birth defects among their children were said to be higher than normal. Vietnamese scientists also reported an unusually high number of birth defects among children of men exposed to the herbicide.

392. Cultural change in animals.
This British tit is sitting on a milk bottle whose foil cap it has pecked open. In the 1930s and 1940s millions of tits in Britain learned how to open milk bottles. This development became perhaps the most famous example of rapid broad-scale change in animal behavior—provoked, in this case, by human industry. It remained an unanswered question whether the cultural shift would have had an effect on the evolution of the British songbirds; the practice of pecking open milk bottles stopped when human beings started putting the bottles in crates.

393. The limits of genetic engineering. Scientists' development of gene-splicing techniques that could genetically alter organisms raised a host of moral, ethical, and public health questions in the 1970s. In many countries the controversy entered the public arena. This 1977 American cartoon reflects the public's growing fears of the results of laboratory research on DNA.

394. Ignorance kills. One of the most significant public health problems to come to the fore in the 1980s was the disease AIDS (acquired immunodeficiency syndrome). Developing a cure, or even effective treatments, for the deadly disease was a difficult task, but scientists and public health officials argued that protection against infection was possible. The virus causing AIDS, it was discovered, could not be spread through casual human contact; use of a condom in sexual intercourse could offer substantial protection. A public armed with knowledge of how the disease spread could, therefore, take measures to protect itself. The advance of science and technology in the twentieth century brought human beings new power to disrupt the biosphere, but it also brought new weapons against the perennial problems of disease, famine, and poverty. The continued search for knowledge and understanding of the natural world, rather than the persistent desire to control it, still offered the best strategy for human survival.

Guide to Further Reading

Twentieth-century biology has produced a vast quantity of research monographs, articles, reports, and the like. Historians and sociologists of science and medicine did not begin analyzing the succession of intellectual and social concerns that mark this immense literature until after mid-century. A number of valuable works have already been written, however, and readers wishing to learn more about the people, discoveries, and issues figuring in this *Album* will find the following list helpful. This selection of English-language sources is a quite personal one. It is but a starting point; many of the publications listed contain useful bibliography for further exploration. In mentioning works I tried to give the date of original publication or of latest revision (when this was known to me).

To learn about the lives and work of individual scientists, the best general, serious source of first resort is the *Dictionary of Scientific Biography*, ed. Charles Coulston Gillispie, 15 vols. plus supplements (New York: Scribners, 1970—), which has helpful bibliographical information.

The place of biology in the growth of twentieth-century science can best be judged by comparing it with the position of physics, chemistry, and medicine in the century's early years. In *A History of the Sciences*, new rev. ed. (New York: Collier, 1962) Stephen F. Mason provides a reliable review of major trends that set the context of biological discoveries early in the century. For such a review of medically oriented research, see Charles Singer and E. Ashworth Underwood, *A Short History of Medicine*, rev. enl. ed. (New York: Oxford University Press, 1962). The broader social effects of medical research are examined in George Rosen's *A History of Public Health* (New York: MD Publications, 1958).

These general histories of science and medicine, however, give scant attention to mid-century discoveries that profoundly altered the face and social impact of biological research. Critical historical analysis of twentieth-century biology as a whole has only just begun. The sole general historical account available is Garland E. Allen's *Life Science in the Twentieth Century* (London: Cambridge University Press, 1978). It and *Biology and the Future of Man*, ed. Philip Handler (New York: Oxford University Press, 1970), outline the major trends associated with biology's spectacular growth. Allen's book examines the rise of experimental biology and the development of its major conceptual foci and features a bibliographical essay surveying recent historical literature. The Handler volume, which grew out of the work of the Committee on Science and Public Policy of the U.S. National Academy of Sciences, gives considerable attention to applied concerns. In its organization it reflects the rethinking of the boundaries between the biological, medical, and agricultural sciences that followed the birth of molecular biology at mid-century. *Biology and the Future of Man* is a rich primary resource revealing much about the complex interplay between biology, physics, and chemistry and about perceived social needs that stimulated the growth of science in the twentieth century. Also useful is Benjamin Dawes's *A Hundred Years of Biology* (New York: Macmillan, 1952). It provides an excellent summary of methodological and conceptual advances as well as a complete bibliography up to mid-century. Only Allen's account, however, supplies the interpretive context for understanding the growth of the vast and successful enterprise called the biological sciences.

The vision and goals of biological investigators early in

the century and the social context of their research are explored in two recent collections: *Physiology in the American Context, 1850–1940* (Bethesda, Md.: American Physiological Society, 1987) and *The American Development of Biology* (Philadelphia: University of Pennsylvania Press, 1988)—the former edited by Gerald L. Geison, the latter by Ronald Rainger, Keith R. Benson, and Jane Maienschein. Individual contributions to these two works supply citations to standard histories of individual disciplines within the natural history and anatomical-physiological traditions. Although focused on American developments, both volumes address the European context of much late-nineteenth-century biology and evaluate the changing significance of experimental analysis within the life sciences at the turn of the twentieth century. They also provide a valuable glimpse into the changing role of the laboratory at the time. The focus and extent of the emerging set of concerns called biology can be seen in *Defining Biology: Lectures From the 1890s*, ed. Jane Maienschein (Cambridge, Mass.: Harvard University Press, 1986).

Nobel laureate François Jacob explains the leading intellectual concerns of twentieth-century biology, emphasizing philosophical and conceptual rather than institutional issues, in *The Logic of Life: A History of Heredity*, trans. Betty E. Spillmann with new preface by author (New York: Pantheon, 1982; orig. French pub. 1970). It is a provocative and readable account that examines the joining of the natural history and anatomical-physiological traditions in studies of reproduction. In *Evolution: The History of an Idea* (Berkeley, Calif.: University of California Press, 1984) Peter J. Bowler explains the subtleties of modern evolution theory. Gordon Rattray Taylor's volume *The Science of Life: A Picture History of Biology* (New York: McGraw-Hill, 1963) briefly reviews major landmark discoveries and is liberally supplied with illustrations, including photographs. In addition to these books, readers seeking surveys of outstanding features of the conceptual growth of biology might also turn to two collections of reprints of classic papers: Mordecai L. Gabriel and Seymour Fogel, eds., *Great Experiments in Biology* (Englewood Cliffs, N.J.: Prentice-Hall, 1955), and Elof A. Carlson, ed., *Modern Biology: Its Conceptual Foundations* (New York: Braziller, 1967).

These intellectual histories are now being supplemented by works that emphasize the role of social and economic forces in the development of scientific ideas—an approach particularly fruitful for understanding the growth of the biological sciences. Several authors have examined, for example, the important influence of eugenic thought in twentieth-century biology. A good starting point here is Daniel J. Kevles, *In the Name of Eugenics: Genetics and the Uses of Human Heredity* (New York: Knopf, 1985). The particular subjective problems inherent in "objective" measurement of biological phenomena are discussed by Stephen Jay Gould in *The Mismeasure of Man* (New York: Norton, 1981).

Biological debates are presented in their broad social context in several good general works that explore the excitement and popular controversy inherent in the concerns of the life sciences in the twentieth century. For a detailed account of the rise of molecular biology, see Horace Freeland Judson, *The Eighth Day of Creation: Makers of the Revolution in Biology* (New York: Simon and Schuster, 1979); for a provocative analysis of the rise of hereditarian and genetic thought in twentieth-century biology, R. C. Lewontin, Steven Rose, and Leon J. Kamin, *Not in Our Genes: Biology, Ideology, and Human Nature* (New York: Pantheon, 1984). Similarly, the focus and accomplishments of medical researchers and their close associations with pharmaceutical interests may be gleaned from Harry F. Dowling's very readable volume *Fighting Infection: Conquests of the Twentieth Century* (Cambridge, Mass.: Harvard University Press, 1977). Roger Lewin's *Bones of Contention: Controversies in the Search for Human Origins* (New York: Simon and Schuster, 1987) reveals the role of evidence, negotiation, and debate in the construction of biological theories, as well as the changing criteria of scientific explanation in the twentieth century. The interface between a popular movement and scientific research is explored by Clive Wood and Beryl Suitters in *The Fight for Acceptance: A History of Contraception* (Aylesbury, England: Medical and Technical Publishing, 1970). Works such as these go beyond the history of ideas to examine how research interests adapt to particular cultural concerns and reflect much more than internal scientific debate and scientific issues.

Several recent biographies deal with the relationship between intellectual development and social context in biological discovery. Kenneth R. Manning gives a sensitive and compelling account of the life of the black American developmental biologist Ernest Everett Just in *Black Apollo of Science* (New York: Oxford University Press, 1983). This fine book vividly portrays the Marine Biological Laboratory at Woods Hole, the Naples Zoological Station, and other leading biological communities of the United States and Europe in the 1910s, 1920s, and 1930s. Evelyn Fox Keller's *A Feeling for the Organism*, (San Francisco: W. H. Freeman, 1983), on the life and work of the geneticist Barbara McClintock, a Nobel laureate, presents a cross section of the American genetics community in the second quarter of the century and the place of a gifted woman scientist within it. For an interesting contrast to the professional and personal experiences of Just and McClintock, see the controversial autobiography by their compatriot James Watson: *The Double Helix: A Personal Account of the Discovery of the Structure of DNA* (New York: Atheneum, 1968). Gunther S. Stent prepared a special edition of *The Double Helix* for the Norton critical editions series (New York: Norton,

1980). This volume includes reviews of Watson's book and many of the original papers by Watson, Francis Crick, Maurice Wilkins, and Rosalind Franklin and encourages the reader to view the process of discovery from several distinct vantage points. Philip J. Pauly presents an interpretation of the rise of reductionist thought in twentieth-century biology in his biography of Jacques Loeb, *Controlling Life: Jacques Loeb and the Engineering Ideal in Biology* (New York: Oxford University Press, 1987).

Several other fine biographical works acquaint the reader with the political and economic forces affecting research and debate in the life sciences. The biologists Lancelot Hogben, J. B. S. Haldane, and Joseph Needham appear in Gary Werskey's *The Visible College: The Collective Biography of British Scientific Socialists of the 1930s* (New York: Holt, Rinehart and Winston, 1979). The scientific sagas underlying the development of insulin and penicillin are portrayed in Michael Bliss, *The Discovery of Insulin* (Chicago: University of Chicago Press, 1982), and Gwyn Macfarlane, *Howard Florey: The Making of a Great Scientist* (New York: Oxford University Press, 1979). These lively, readable narratives link larger institutional, political, and economic interests with the individual careers of Frederick Banting, Charles Best, J. J. R. Macleod, and J. B. Collip in the case of insulin and Alexander Fleming, Howard Florey, and Ernst Chain in the case of penicillin. Zhores A. Medvedev describes Lysenko's career and his position within Soviet science in *The Rise and Fall of T. D. Lysenko*, trans. I. Michael Lerner (New York: Columbia University Press, 1969). James Reed well articulates the link between public controversies and research directions within science in *From Private Vice to Public Virtue: The Birth Control Movement and American Society Since 1830* (New York: Basic Books, 1978). The central figures of nurse-activist Margaret Sanger, physician-gynecologist Robert Latou Dickinson, physician-entrepreneur Clarence J. Gamble, and biologist Gregory Pincus illustrate the complicated social and scientific interests behind the development of the oral contraceptive pill.

Conceptual and methodological issues in twentieth-century biology are closely interrelated and as yet incompletely understood. The importance of methodological advances and debates for the discovery and understanding of microorganisms is illustrated in histories of individual fields such as that by A. P. Waterson and Lise Wilkinson, *An Introduction to the History of Virology* (New York: Cambridge University Press, 1978). John Farley's final chapters in his volume *The Spontaneous Generation Controversy From Descartes to Oparin* (Baltimore: The Johns Hopkins University Press, 1977) focus attention on the changing definitions of living and nonliving matter as the twentieth century progressed but before DNA became the primary focus of biological

attention. Erwin Schrödinger's little book *What Is Life? The Physical Aspect of the Living Cell* (London: Cambridge University Press, 1944) shows the physicists' approach to this problem and the beginning of the intellectual revolution provoked by the discovery of the structure of DNA.

Readers can explore the manifold social and economic consequences of twentieth-century biological research in a variety of works. The beneficial effects of improved agricultural practices are surveyed in E. C. Stakman, Richard Bradfield, and Paul C. Mangelsdorf, *Campaigns Against Hunger* (Cambridge, Mass.: Harvard University Press, 1967). Critiques of research emphases in the later decades abound. These include Rachel Carson's classic *Silent Spring* (Boston: Houghton-Mifflin, 1962), as well as Barry Commoner, *Science and Survival* (New York: Viking, 1966), Paul Ehrlich, *The Population Bomb*, 2nd ed. (New York: Ballantine, 1971), and Peter Singer, *Animal Liberation: A New Ethics for Our Treatment of Animals* (New York: New York Review of Books, 1975). A provocative essay from earlier in the century voicing similar concerns is, of course, Aldous Huxley's novel *Brave New World* (Garden City, N.Y.: Doubleday, 1932).

The growth of "big science" is an important theme, especially as biological research has become a significant contributor to the economy and as science policy has become an important responsibility of statecraft. Recent conceptions of progress in the biological sciences and strategies for wedding scientific research to national prosperity are explored in the U.S. National Research Council's *Outlook for Science and Technology: The Next Five Years* (San Francisco: W. H. Freeman, 1982). Special sections on demography, human and plant disease, nutrition, natural resources, waste management, and new technologies attest to continued expectations for economic and social progress through basic research. Derek J. de Solla Price's *Little Science, Big Science* (New York: Columbia University Press, 1963) helped create a new science of science and stimulated critical analysis of how science actually grows.

Readers wishing to pursue the changing concerns of biology in the twentieth century should not neglect the series of readings from *Scientific American* published by W. H. Freeman. For the biographies and Nobel lectures of Nobel laureates, see the series *Nobel Lectures in Physiology-Medicine* (New York: Elsevier, 1964—). Anyone interested in recent scholarship in the history of the biological sciences should of course take note of the major journals. These include *Journal of the History of Biology, History and Philosophy of the Life Sciences, Bulletin of the History of Medicine, Journal of the History of Medicine and Allied Sciences, Medical History, Journal of the History of Behavioral Sciences, Perspectives in Biology and Medicine, Social Studies of Science, Isis, Osiris, Gesnerus,* and *British Journal for the History of Science.*

Picture Sources and Credits

All possible care has been taken to trace the ownership of each illustration and to make full acknowledgment for its use. If any errors or omissions have accidentally occurred, they will be corrected in subsequent editions provided notification is sent to the publisher.

1. James D. Watson, *The Double Helix*, New York, Atheneum, 1968, p. 215; © 1968 by James D. Watson. 2. Neg. No. 327725. Courtesy Department of Library Services, American Museum of Natural History. 3. By permission Trustees, British Museum (Natural History). 4. Neg. No. 327740. Courtesy Department of Library Services, American Museum of Natural History. 5. Neg. No. 327733. Courtesy Department of Library Services, American Museum of Natural History. 6. Brown Brothers. 7. Institut Pasteur, Paris. 8. Institut Pasteur, Paris. 9. The Mansell Collection. 10. Hoechst AG. 11. Hoechst AG; from a drawing by von Lüder. 12. Chicago Historical Society, IBHi-13686. 13. Wellcome Institute for the History of Medicine Library, London. 14. Wellcome Institute for the History of Medicine Library, London. 15. Station Biologique de Roscoff. 16. Institut Pasteur, Paris. 17. *Methods and Problems of Medical Education*, ser. 3, New York, Rockefeller Foundation, 1925. 18. *Methods and Problems of Medical Education*, ser. 3, New York, Rockefeller Foundation, 1925. 19. University of Bern. 20. *Methods and Problems of Medical Education*, ser. 10, New York, Rockefeller Foundation, 1928. 21. Smith College Archives, Smith College. 22. Katherine Elizabeth McClellan, photographer; Smith College Archives, Smith College. 23. William F. Ganong, *A Laboratory Course in Plant Physiology*, 2nd ed., New York, 1908, frontispiece. 24. The Beckman Center for the History of Chemistry, University of Pennsylvania. 25. Archives Photographiques/Musée Pasteur. 26. *Methods and Problems of Medical Education*, ser. 3, New York, Rockefeller Foundation, 1925. 27. *Methods and Problems of Medical Education*, ser. 3, New York, Rockefeller Foundation, 1925. 28. Patrice Boussel, Library of the City of Paris. 29. Patrice Boussel, Library of the City of Paris. 30. Marine Biological Laboratory. 31. Collection, Garland Allen, Washington University, St. Louis. 32. Marine Biological Laboratory. 33. Stazione Zoologica "Antonio Dohrn" di Napoli. 34. Stazione Zoologica "Antonio Dohrn" di Napoli. 35. John C. Bugher Collection, Rockefeller Archive Center. 36. John C. Bugher Collection, Rockefeller Archive Center. 37. Archives Photographiques/Musée Pasteur. 38. Max-Planck-Gesselschaft zur Förderung der Wissenschaften, E.V., Berlin. 39. National Institute for Medical Research, Medical Research Council, London. 40. Rothamsted Experimental Station. 41. U.S. Department of Agriculture. 42. Station Biologique de Roscoff. 43. Jack Calvin, Pacific Grove, California. 44. Station Biologique de Roscoff. 45. National Aeronautics and Space Administration. 46. UPI/Bettmann Newsphotos. 47. National Aeronautics and Space Administration. 48. Hoechst AG. 49. Sprague Institute. 50. The Jackson Laboratory. 51. Dr. Lee Ehrman, State University of New York, Purchase; from Irwin H. Herskowitz, *Principles of Genetics*, New York, Macmillan, 1973. 52. University of Texas, Austin. 53. Ross Institute of Tropical Hygiene; from Carlos Alvarado and L. J. Bruce-Chwatt, "Malaria," *Scientific American*, May 1962, p. 92. 54. Theodore C. Ruch and John F. Fulton, *Medical Physiology and Biophysics*, 18th ed., © 1960, W. B. Saunders Company, p. 1170, fig. 612. 55. University of California. 56. C. A. Knight, *Cold Spring Harbor Symposia*, vol. 12, 1947, pp. 115–121. 57. Dr. Howard Jones, Berkeley, California. 58. Roy de Carava; from F. C. Steward, "The Control of Growth in Plant Cells," *Scientific American*, Oct. 1963, p. 105. 59. School of Public Health, University of California, Berkeley. 60. Squibb Pharmaceutical Company. 61. *Pravda*, Nov. 13, 1957. 62. U.S. Navy. 63. © 1986 by Dr. Seuss and A.

S. Geisel, from *You're Only Old Once*, New York, Random House. **64.** Siemens AB, FRD. **65.** L. Marton Collection, Archive Center/Smithsonian Institution. **66.** © 1984 Jonathan Levine. **67.** University of California, Berkeley. **68.** American Institute of Physics, Niels Bohr Library; from S. Bradbury, *Evolution of the Microscope*, Oxford, Pergamon Press, 1967, fig. 8.13. **69.** Courtesy Jeffery L. Barker, M.D., Chief, Laboratory of Neurophysiology, National Institute of Neurological and Communicative Disorders and Stroke, National Institutes of Health. **70.** Francis A. Countway Library of Medicine, Harvard Medical Library. **71.** National Library of Medicine. **72.** *Cancer*, vol. 5, 1952, p. 1. **73.** V. C. Littau, V. G. Allfrey, J. H. Frenster, and A. E. Mirsky, *Proceedings of the National Academy of Science*, vol. 52, 1964, p. 97. **74.** J. D. Bernal, Department of Crystallography, University of London. **75.** *Seminars in Roentgenology*, vol. 12, 1977, p. 11. **76.** David Grunfeld/New York State College of Veterinary Medicine at Cornell. **77.** University College, London. **78.** Smith College Archives. **79.** U.S. Navy Photograph. **80.** Neg. No. 2A5461, photo by Logan. Courtesy Department of Library Services, American Museum of Natural History. **81.** American Physiological Society. **82.** Julius H. Comroe, Jr., *Exploring the Heart*, New York, Norton, 1983, p. 281. **83.** University of California, Berkeley. **84.** University of Cambridge Library. **85.** G. R. Taylor, *The Science of Life*, New York, McGraw-Hill, 1963, p. 330. **86.** Imperial Chemical Industries. **87.** George W. Gray/Rockefeller University. **88.** David Linton; from William H. Stein and Stanford Moore, "The Chemical Structure of Proteins," *Scientific American*, Feb. 1961, p. 86. **89.** Roy de Carava; from P. C. Trexler, "Germ-Free Isolators," *Scientific American*, July 1964, p. 80. **90.** Rockefeller Archive Center. **91.** Don Siegel, Harvard University, and Robert Fleischaker, Massachusetts Institute of Technology; from Herman J. Phaff, "Industrial Microorganisms," *Scientific American*, Sept. 1981, p. 86. **92.** Smithsonian Institution Neg. No. 44682. **93.** Neg. No. 315147, photo by Charles H. Coles. Courtesy Department of Library Services, American Museum of Natural History. **94.** National Park Service, U.S. Department of the Interior. **95.** Zoological Institute, U.S.S.R. Academy of Sciences, Leningrad. **96.** Zoological Institute, U.S.S.R. Academy of Sciences, Leningrad. **97.** Neg. No. 116839, photo by W. Thompson. Courtesy Department of Library Services, American Museum of Natural History. **98.** University of Arizona, Department of Geosciences. **99.** Neg. No. 326764. Courtesy Department of Library Services, American Museum of Natural History. **100.** Texas Memorial Museum, Vertebrate Paleontology Laboratory, University of Texas, Austin. **101.** Neg. No. 333451, photo by Anderson. Courtesy Department of Library Services, American Museum of Natural History. **102.** Neg. No. 46501, photo by Stratford. Courtesy Department of Library Services, American Museum of Natural History. **103.** Neg. No. 311515, photo by E. M. Fulda. Courtesy Department of Library Services, American Museum of Natural History. **104.** Neg. No. 35331, photo by Thomson. Courtesy Department of Library Services, American Museum of Natural History. **105.** Neg. No. 314804, photo by Julius Kirschner. Courtesy Department of Library Services, American Museum of Natural History. **106.** American Museum of Natural History. **107.** B. I. Balinsky, *Introduction to Embryology*, Philadelphia, W. B. Saunders, 1960, fig. 155. **108.** Station Biologique de Roscoff. **109.** Marine Biological Laboratory. **110.** Friedrich Baltzer, *Theodor Boveri: Life and Work of a Great Biologist*, Berkeley, University of California Press, 1967, p. 61. **111.** G. R. Taylor, *The Science of Life*, New York, McGraw-Hill, 1963, p. 256. **112.** Charles Sedgwick Minot, "The Problem of Age, Growth, and Death," *Popular Science Monthly*, vol. LXXI, June-Dec. 1907, p. 364. **113.** W. J. Robbins et al., *Growth*, New Haven, Yale University Press, 1928, p. 119. **114.** B. I. Balinsky, *An Introduction to Embryology*, Philadelphia, W. B. Saunders, 1960, fig. 55. **115.** Courtesy of B. I. Balinsky and W. B. Saunders & Company. **116.** B. I. Balinsky, *An Introduction to Embryology*, Philadelphia, W. B. Saunders, 1960, fig. 190. **117.** Courtesy of B. I. Balinsky and W. B. Saunders & Company. **118.** T. C. Kramer, "Cinemicrographic Studies of Rabbit Ovulation," *The Anatomical Records*, vol. 63, 1935, pp. 239–245, plate p. 245. **119.** John C. Finerty and E. V. Cowdry, *A Textbook of Histology*, 5th ed., Philadelphia, Lea & Febiger, 1962, fig. 321. **120.** Mia Tegnor, Scripps Institution of Oceanography; from David Epel, "The Program of Fertilization," *Scientific American*, Nov. 1977, p. 128. **121.** Arthur Lentz Colwin and Laura Hunter Colwin, in Society for the Study of Development and Growth, *Cellular Membranes in Development*, New York, Academic Press, 1964. **122.** Collection, Garland Allen, Washington University, St. Louis; © Ralph Crane/Black Star. **123.** Medelianum Musei Moraviae, Brno, Czechoslovakia, Office of the Director, Mr. V. Orel. **124.** Medelianum Musei Moraviae, Brno, Czechoslovakia, Office of the Director, Mr. V. Orel. **125.** Rijksmuseum voor de Geschjedenis der Natuurweten schappen. **126.** U.S. Department of Agriculture. **127.** Collection, Garland Allen, Washington University, St. Louis; from T. H. Morgan et al, *The Mechanism of Medelian Heredity*, New York, Holt, 1915, fig. 14. **128.** J. C. Ewart, *The Penycuik Experiments*, London, Adam and Charles Black, 1899, fig. 3. **129.** Cold Spring Harbor Laboratory Research Library Archives. **130.** Bruce Wallace, *Chromosomes, Giant Molecules and Evolution*, New York, Norton, 1966. **131.** Collection, Garland Allen, Washington University, St. Louis; from T. H. Morgan, "An Attempt to Analyze the Constitution of the Chromosomes on the Basis of Sex-Limited Inheritance in *Drosophila*," *Journal of Experimental Zoology*, vol. 11, 1911, plate I. **132.** Courtesy of the Archives, California Institute of Technology. **133.** T. S. Painter, "A New Method for the Study of Chromosome Rearrangements and the Plotting of Chromosome Maps," *Science*, Dec. 22, 1933, p. 586. **134.** T. S. Painter, "Chromosomes in Drosophila," *Genetics*, vol. 19, 1934, p. 179. **135.** After Victor A. McKusick, *Human Genetics*, © 1964; reproduced by permission of Prentice-Hall, Inc. **136.** Ernest H. Starling, *Principles of Human Physiology*, Philadelphia, Lea & Febiger, 1912, p. 14. **137.** Dr. Muriel T. Davisson, The Jackson Laboratory. **138.** Cold Spring Harbor Research Library Archives. **139.** Marcus Rhoades, Department of Biology, Indiana University,

Bloomington. **140.** Brookhaven National Laboratory. **141.** Brookhaven National Laboratory. **142.** National Portrait Gallery, London. **143.** Charles Elton, *Animal Ecology*, New York, Macmillan, 1927, opposite p. 40, plate V (top). **144.** Charles Elton, *Animal Ecology*, New York, Macmillan, 1927, opposite p. 40, plate V (bottom). **145.** Victor E. Shelford, *Animal Communities in Temperate America*, © 1913 University of Chicago Press. Used by permission. **146.** E. C. Williams, "An Ecological Study of the Floor Fauna of the Panama Rain Forest," *Bulletin of the Chicago Academy of Science*, vol. 6, 1941, pp. 63–124. **147.** G. F. Gause, *The Struggle for Existence*, Baltimore, Williams & Wilkins Company, 1934. **148.** Bunj Tagawa; from Bert Bolin, "The Carbon Cycle," *Scientific American*, Sept. 1970, p. 126. **149.** U.S. Department of Agriculture. **150.** U.S. Department of Agriculture, *After a Hundred Years: The Yearbook of Agriculture 1962*, p. 156. **151.** A. J. Anderson, Commonwealth Scientific & Industrial Research Organization, Plant Industry Division; from Anderson and Underwood, "Trace-Element Deserts," *Scientific American*, Jan. 1959, p. 98. **152.** A. J. Anderson, Commonwealth Scientific & Industrial Research Organization, Plant Industry Division; from Anderson and Underwood, "Trace-Element Deserts," *Scientific American*, Jan. 1959, pp. 98–99. **153.** A. J. Anderson, Commonwealth Scientific & Industrial Research Organization, Plant Industry Division; from Anderson and Underwood, "Trace-Element Deserts," *Scientific American*, Jan. 1959, p. 99. **154.** Donald H. Marx, Southeastern Forest Experimental Station; from W. J. Brill, "Soil Microbiology," *Scientific American*, Sept. 1981, p. 208. **155.** U.S. Department of Agriculture Forest Service. **156.** Richard Pardo, American Forestry Association; from S. H. Spurr, "Silviculture," *Scientific American*, Feb. 1979, p. 77. **157.** J. L. B. Smith Institute of Ichthyology, Rhodes University, South Africa. **158.** T. H. Morgan, *A Critique of the History of Evolution*, Princeton, Princeton University Press, 1916, fig. 95. **159.** Joseph Barrell et al., *The Evolution of the Earth and Its Inhabitants*, New Haven, Yale University Press, 1923, fig. 14. **160.** J. A. Moore, *Biological Science: An Inquiry Into Life*, New York, Harcourt Brace. **161.** Reproduced by courtesy of Cambridge University Press. **162.** Professor H. B. D. Kettlewell, University of Oxford. **163.** Professor H. B. D. Kettlewell, University of Oxford. **164.** Gerhard Heberer and Franz Schwanitz, *Hundert Jahre Evolutionsforschung*, Stuttgart, FRD, Gustav Fischer Verlag, 1960, p. 199. **165.** British Museum (Natural History). **166.** G. Ledyard Stebbins, *Processes of Organic Evolution*, 3rd ed., Englewood Cliffs, N.J., Prentice-Hall, 1977, fig. 1-1. **167.** R. H. Whittaker, *Science*, vol. 163, Jan. 10, 1969, pp. 150–160. **168.** Elso Barghoorn; from A. Lee McAlester, *The History of Life*, 2nd ed., Englewood Cliffs, N.J., Prentice-Hall, 1977, fig. 1-9, p. 17. **169.** Photograph courtesy of Professor M. F. Glaessner, University of Adelaide, South Australia. **170.** Geological Survey of Canada. **171.** A. P. Waterson and Lise Wilkinson, *An Introduction to the History of Virology*, New York, Cambridge University Press, 1978; originally published by Ivanovsky in "Über die Mosaikkrankheit der Tabakspflanze," *Zeitschrift für Pflanzenkrankheiten*, vol. 13, 1903, pp. 1–41. **172.** Friedrich Loeffler, "Weitere Untersuchungen über die Beizung und Farbung der Geisseln bei den Bakterien," *Zentralblatt für Bakteriologie*, vol. 7, 1890, pp. 625–639. **173.** American Physiological Society. **174.** Neg. No. 312582, photo by H. S. Rice and I. Datcher. Courtesy Department of Library Services, American Museum of Natural History. **175.** British Empire Cancer Campaign. **176.** Sheldon J. Segal and G. P. Talwar, Rockefeller Institute; from Eric G. Davidson, "Hormones and Genes," *Scientific American*, June 1965, reprinted in *The Molecular Basis of Life*, San Francisco, W. H. Freeman, 1967. **177.** Irwin H. Herskowitz, *Principles of Genetics*, New York, Macmillan, 1973, p. 6, fig. 1-4. **178.** George Palade, Rockefeller Institute; from H. F. Judson, *The Eighth Day of Creation*, New York, Simon and Schuster, 1979. **179.** H. E. Huxley, "The Contraction of Muscle," *Scientific American*, Nov. 1958, reprinted in *The Living Cell*, San Francisco, W. H. Freeman, 1965. **180.** (left) Peter Satir, Department of Anatomy, Albert Einstein College of Medicine, Bronx, New York; from William A. Jensen and Roderick B. Park, *Cell Ultrastructure*, Belmont, Calif., Wadsworth Publishing Company, 1967, fig. 10-2. **180.** (right) David Ringo, Yale University School of Medicine; from William A. Jensen and Roderick B. Park, *Cell Ultrastructure*, Belmont, Calif., Wadsworth Publishing Company, 1967, fig. 10-3. **181.** D. E. Bradley, "Electron Microscopy: Botanical Applications," George L. Clark, ed., *The Encyclopedia of Microscopy*, New York, Reinhold, 1961, p. 83. **182.** R. Falk, Hebrew University, Jerusalem, Israel. **183.** H. R. Duncker, Zentrum für Anatomie und Cytobiologie, Giessen, FRD; from Knut Schmidt-Nielsen, *Animal Physiology*, New York, Cambridge University Press, 1975, p. 56. **184.** Roger D. Meicenheimer, Department of Botany, Miami University, Oxford, Ohio. **185.** Emil Bernstein and Elia Kairinen, Gillette Research Institute; first published on the cover of *Science*, vol. 173, Aug. 27, 1971, copyright 1971 by the American Association for the Advancement of Science. **186.** Professor A. C. Nelson, University of Washington. **187.** R. B. Park. **188.** Stanley Cohen, Department of Genetics, Stanford University School of Medicine. **189.** U.S. Department of Agriculture. **190.** *Hygeia*, vol. 1, 1923, p. 519, a publication of the American Medical Association. **191.** Charles Herbert Best and Norman Burke Taylor, *The Physiological Basis for Medical Practice*, 5th ed., Baltimore, Williams and Wilkins, 1950, p. 613, fig. 266. **192.** American Physiological Society. **193.** American Physiological Society. **194.** Mayo Foundation, Rochester, Minn.; reproduced originally by E. H. Wood, E. H. Lambert, E. J. Baldes, and C. F. Code, "Effects of Acceleration in Relation to Aviation," *Federation Proceedings*, vol. 3, 1946, pp. 327–444. **195.** Vance Tucker, Department of Zoology, Duke University, Durham, N.C. **196.** Clement A. Smith; from Clement A. Smith, "The First Breath," *Scientific American*, Oct. 1963, p. 34. **197.** University of Michigan Photographic Services. **198.** Francis A. Countway Library of Medicine, Harvard University. **199.** American Physiological Society. **200.** Smithsonian Institution Neg. No. 76-17266. **201.** National Library of Medicine. **202.** *Methods and Problems of Medical Education*, ser. 5, New York, Rockefeller Foundation, 1924. **203.** Earl H. Wood,

Biodynamic Research Unit, Mayo Medical School, Rochester, Minn.; from *Annals of Biomedical Engineering*, 1978, p. 755. **204.** Jon Hoffmann/Photographers International. **205.** Courtesy of Gould, Inc., Brush Instruments Division, Cleveland, Ohio. **206.** Honeywell Biomedical Products, Electronic Products, Colorado. **207.** National Aeronautics and Space Administration; from Mae Mills Link, *Space Medicine in Project Mercury* [NASA SP-4003], 1965, fig. 1. **208.** Bettmann Archive. **209.** University of Michigan Photographic Services. **210.** Judith P. Swazey, *Reflexes and Motor Integration*, Cambridge, Mass., Harvard University Press, 1969, fig. 6. **211.** W. F. Evans, *Anatomy and Physiology, the Basic Principles*, Prentice-Hall, N.J., 1971. **212.** *Journal of Experimental Zoology*, vol. 23, 1917, p. 426. **213.** The Mayo Clinic. **214.** The Mayo Clinic. **215.** Michael Bliss, *The Discovery of Insulin*, Chicago, University of Chicago Press, 1982; from Fisher Library, University of Toronto, the F. G. Banting Papers. **216.** University of Toronto Archives. **217.** University of Utrecht. **218.** Margaret S. Livingstone, Harvard Medical School. **219.** Averett S. Tombes, *An Introduction to Invertebrate Endocrinology*, New York, Academic Press, 1970, fig. 1-4; after B. E. Frye, *Hormonal Control in Vertebrates*, Macmillan, New York, 1967. **220.** University of Wisconsin College of Agriculture. **221.** E. F. Smith Memorial Collection, The Beckman Center for the History of Chemistry, University of Pennsylvania. **222.** E. F. Smith Memorial Collection, The Beckman Center for the History of Chemistry, University of Pennsylvania; 1914 Arthur H. Thomas Company Catalogue. **223.** The Mayo Clinic. **224.** The Mayo Clinic. **225.** The Mayo Clinic. **226.** Max-Planck-Gesselschaft zur Förderung der Wissenschaften, E.V., Berlin. **227.** Max-Planck-Gesselschaft zur Förderung der Wissenschaften, E.V., Berlin. **228.** Max-Planck-Gesselschaft zur Förderung der Wissenschaften, E.V., Berlin. **229.** Brookhaven National Laboratory. **230.** Jon Brennis; from Arnon, "The Role of Light in Photosynthesis," *Scientific American*, Nov. 1960, pp. 111–112. **231.** L. K. Shumway, Washington State University. **232.** Sol Mednick; from Marshall W. Nirenberg, "The Genetic Code: II," *Scientific American*, March 1963, pp. 80–81. **233.** A. K. Kleinschmidt, Universitat Ulm, Abteilung Virologie, Ulm, Germany. **234.** King's College London, University of London, Department of Biophysics. **235.** King's College London, University of London, Department of Biophysics. **236.** Courtesy of Vittorio Luzzati. **237.** Reprinted by permission from *Nature*, vol. 171, p. 737; copyright © 1953 Macmillan Magazines Ltd. **238.** Reprinted by permission from *Nature*, vol. 227, pp. 561–562; copyright © 1970 Macmillan Magazines Ltd. **239.** Gunther S. Stent, *Molecular Biology of Bacterial Viruses*, San Francisco, W. H. Freeman, 1963. **240.** *Paris-Match.* **241.** Gunther S. Stent, in *The Neurosciences*, New York, Rockefeller University Press, 1967; reproduced in G. S. Stent, *The Coming of the Golden Age*, Garden City, N.Y., Natural History Press, 1969. **242.** O. L. Miller, Jr., and Barbara R. Beatty, "Portrait of a Gene," *Journal of Cellular Physiology*, vol. 74, sup. 1, 1969, pp. 225–232. **243.** California Institute of Technology. **244.** Oswald T. Avery, Colin M. MacLeod, and Maclyn McCarty, "Induction of Transformation by a Desoxyribonucleic Acid Fraction Isolated From Pneumococcus Type III," *Journal of Experimental Medicine*, vol. 79, 1944, plate 1, opposite p. 158; photo by Joseph B. Haulenbeek. **245.** Thomas F. Anderson, Fox Chase Cancer Center, Philadelphia, Pennsyvania. **246.** Thomas F. Anderson, Fox Chase Cancer Center, Philadelphia, Pennsylvania. **247.** Edouard Kellenberger, University of Geneva; from Jacob and Wollman, "Viruses and Genes," *Scientific American*, June 1961, p. 100. **248.** Edouard Kellenberger, University of Geneva; from Jacob and Wollman, "Viruses and Genes," *Scientific American*, June 1961, p. 100. **249.** © Keystone/The Image of Works. **250.** Melvin Calvin, University of California, Berkeley. **251.** George Wald, "The Origin of Life," *Scientific American*, August 1954. **252.** Stanley Miller, Department of Chemistry, University of California, San Diego. **253.** A. L. Oparin; from R. E. Dickerson, "Chemical Evolution and the Origin of Life," *Scientific American*, September 1978, p. 83. **254.** Sidney W. Fox; from R. E. Dickerson, "Chemical Evolution and the Origin of Life," *Scientific American*, September 1978, p. 83. **255.** James F. Shepard, Kansas State University; from James F. Shepard, "The Regeneration of Potato Plants From Leaf-Cell Protoplasts," *Scientific American*, May 1982, p. 155. **256.** James F. Shepard, Kansas State University; from James F. Shepard, "The Regeneration of Potato Plants From Leaf-Cell Protoplasts," *Scientific American*, May 1982, p. 158. **257.** R. Briggs and T. J. King, "Transmutation of Living Nuclei From Bastula Cells Into Enucleated Frogs' Eggs," *Proceedings of the National Academy of Sciences*, vol. 38, pp. 455–463 (fig. B). **258.** Peter C. Huppe, The Jackson Lab/Cell Magazine; *Science*, vol. 211, pp. 351–358, cover, Jan. 23, 1981. **259.** William Marin, Jr./Brookhaven National Laboratory; *Science*, July 30, 1976. **260.** National Academy of Sciences. **261.** U.S. Department of State. **262.** Monsanto Enviro-Chem Systems; used by permission. **263.** William K. Gregory, *Our Face From Fish to Man*, New York, Putnam, 1929, frontispiece. **264.** Winterton C. Curtis, *Fundamentalism vs. Evolution at Dayton, Tennessee*, Falmouth, Mass., The Falmouth Enterprise, 1956. **265.** Winterton C. Curtis, *Fundamentalism vs. Evolution at Dayton, Tennessee*, Falmouth, Mass., The Falmouth Enterprise, 1956. **266.** UPI/Bettmann Newsphotos. **267.** UPI/Bettmann Newsphotos. **268.** A London Journalist [Newman Watts], *Why Be An Ape—?*, London, Marshall, Morgan & Scott, 1942. **269.** Geological Society of London; painting by John Cooke, 1915. **270.** (skeletons) "Fossil Man," *Nature*, vol. 91, no. 2287, August 28, 1913, p. 663. **270.** (skull) Musée de l'Homme, Paris. **271.** E. A. Hooten, *Up From the Ape*, New York, Macmillan, 1937, fig. 57. **272.** Henry Fairfield Osborn, "The Discovery of Tertiary Man," *Nature*, vol. 125, no. 3141, January 11, 1930, p. 54; from the American Museum of Natural History. **273.** Sherwood L. Washburn, *Scientific American*, September 1960. **274.** The National Museums of Kenya, Nairobi, Kenya. **275.** UPI/Bettmann Newsphotos. **276.** Neg. No. 24485. Courtesy Department of Library Services, American Museum of Natural History. **277.** Jean-Loup Charmet. **278.** Edmund Engelman. **279.** Robert Mearns Yerkes and Ada Watterson Yerkes, *The Great*

Apes, New Haven, Yale University Press, 1929; photographs made in the laboratory of Mrs. Nadie Kohts, Moscow. **280.** © Dr. Ronald H. Kohn/The Gorilla Foundation. **281.** AP/Wide World Photos. **282.** Konrad Lorenz, *On Aggression*, New York, Marjorie Kerr Wilson, trans., Harcourt, Brace & World, 1966, p. 184, fig. 5. **283.** Thomas McAvoy/LIFE Magazine © 1955 Time, Inc. **284.** B. F. Skinner, Harvard University. **285.** Neg. No. 2A6053, photo by Logan. Courtesy Department of Library Services, American Museum of Natural History. **286.** University of Wisconsin Primate Laboratory. **287.** E. O. Wilson, Museum of Comparative Zoology, Harvard University. **288.** Courtesy John B. Calhoun, Ph.D./ Calhoun Collection of the American Heritage Center, University of Wyoming, Laramie. **289.** Courtesy John B. Calhoun, Ph.D./Calhoun Collection of the American Heritage Center, University of Wyoming, Laramie. **290.** Dr. John B. Calhoun, National Institutes of Health. **291.** Institut Pasteur, Paris. **292.** Hoechst AG. **293.** Social Welfare History Archives, University of Minnesota, from the American Social Health Association Records. **294.** Bayer AG, Leverkusen, FRG. **295.** National Library of Medicine/History of Medicine Archive. **296.** Wesley W. Spink, *Sulfanilamide and Its Related Compounds in General Practice*, 2nd ed., Chicago, Year Book Publishers, Inc., 1942, p. 45. **297.** Merck Sharpe & Dohme. **298.** Squibb Pharmaceutical Company. **299.** Eli Lilly & Company. **300.** E. R. Squibb & Sons, Inc. **301.** Courtesy Pfizer, Inc. **302.** © Charles Harbutt/Archive Pictures. **303.** Eli Lilly & Company. **304.** John Freeman. **305.** John Tellick, London. **306.** Courtesy, United Nations Children's Fund USA (UNICEF)/Avild Vollan, no. 8717/81. **307.** Sir Frank Macfarlane Burnet and David O. White, *Natural History of Infectious Diseases*, 4th ed., New London, Cambridge University Press, 1972. **308.** John C. Bugher Collection, Rockefeller Archive Center. **309.** John C. Bugher Collection, Rockefeller Arvhive Center. **310.** John C. Bugher Collection, Rockefeller Archive Center. **311.** National Library of Medicine Historical Collection/PAR/NYC Photo. **312.** Elizabeth Harris, Click/Chicago. **313.** *Hygeia*, vol. 16, 1938, p. 699, a publication of the American Medical Association. **314.** World Health Organization. **315.** Centers for Disease Control. **316.** Donald A. Henderson, "The Eradication of Smallpox," *Scientific American*, October 1976. **317.** International Planned Parenthood Federation. **318.** Planned Parenthood Federation of America. **319.** Planned Parenthood Federation of America. **320.** Library of Congress. **321.** Planned Parenthood World Federation. **322.** Neg. No. 33724, photo by P. Hollembeak. Courtesy Department of Library Services, American Museum of Natural History. **323.** *The Aims of Eugenics*, London, The Eugenics Society, n.d., frontispiece. **324.** Alan S. Parkes, *Off-Beat Biologist*, Cambridge, The Galton Foundation, 1985, p. 232. **325.** Planned Parenthood Federation of America. **326.** International Planned Parenthood Federation. **327.** International Planned Parenthood Federation. **328.** Robert E. Murowchick/Photo Researchers. **329.** Robert E. Murowchick/Photo Researchers. **330.** Robert E. Murowchick/Photo Researchers. **331.** Robert E. Murowchick/ Photo Researchers. **332.** U.S. Department of Agriculture. **333.** Garland Allen/Washington University; from Ernst Haeck Hause, Jena, DDR. **334.** Empire Cotton Growing Corporation. **335.** U.S. Department of Agriculture. **336.** Henry A. Wallace and William L. Brown, *Corn and Its Early Fathers*, East Lansing, Mich., Michigan State University Press, 1956, fig. 7. **337.** Henry A. Wallance and William L. Brown, *Corn and Its Early Fathers*, East Lansing, Mich., Michican State University Press, 1956, fig. 8. **338.** U.S. Department of Agriculture, *After a Hundred Years: The Yearbook of Agriculture 1962*, p. 144. **339.** Agricultural Research Service, U.S. Department of Agriculture. **340.** U.S. Department of Agriculture. **341.** Charlton Photos © 1982. **342.** Ball Corporation. **343.** U.S. Department of Agriculture, *After a Hundred Years: The Yearbook of Agriculture 1962*, p. 43. **344.** New York State College of Agriculture and Life Sciences/Cornell University, photo by Don Albern. **345.** George Seidel, Animal Reproduction Laboratory, Colorado State University. **346.** Animal Reproduction Laboratory, Colorado State University. **347.** The Museum of Modern Art/Film Stills Archive. **348.** *The Illustrated London News*, Sept. 15, 1928, p. 453. **349.** *Galaxy*, Sept. 1954, cover. **350.** The Children's Hospital, Boston. **351.** Ali Seireg, University of Wisconsin. **352.** Medtronic. **353.** World Health Organization. **354.** J. J. Abel, L. G. Rowntree, and B. B. Turner, "On the Removal of Diffusible Substances From the Circulating Blood of Living Animals by Dialysis," *Journal of Pharmacology and Experimental Therapeutics*, vol. 5, 1914, pp. 275–316. **355.** J. J. Abel, L. G. Rowntree, and B. B. Turner, "On the Removal of Diffusible Substances From the Circulating Blood of Living Animals by Dialysis," *Journal of Pharmacology and Experimental Therapeutics*, vol. 5, 1914, pp. 275–316. **356.** Bob Walsh; Alan E. Nourse and the Editors of Time-Life Books, *The Body*, Alexandria, Va., Time-Life Books, 1980, pp. 128–129. **357.** (top and bottom) A. Carrel and C. C. Guthrie, "Uniterminal and Biterminal Venous Transplantations," *Surgery, Gynecology & Obstetrics*, vol. 2, 1906, pp. 266–286. **357.** (middle) A. Carrel, "Results of the Permanent Intubation of the Thoracic Aorta," *Surgery, Gynecology & Obstetrics*, vol. 15, 1912, pp. 245–248. **358.** Courtesy Loma Linda University, California. **359.** Ralph Morse, LIFE Magazine © Time, Inc. **360.** Dr. Willem J. Kolff, University of Utah. **361.** Courtesy University of Utah Medical Center. **362.** Stanford University Medical Center. **363.** Eric Mose; from Homer W. Smith, "The Kidney," *Scientific American*, Jan. 1953, pp. 46–47. **364.** L. J. Henderson, *Blood: A Study in General Physiology*, New Haven, Yale University Press, 1928, p. 98. **365.** Courtesy Veterans Administration Hospital, Sepulveda, Calif. **366.** Photomicrograph by Moses Kunitz; from H. Neurath, "Protein-Digesting Enzymes," *Scientific American*, Dec. 1964, p. 78. **367.** Photomicrograph by Moses Kunitz; from H. Neurath, "Protein-Digesting Enzymes," *Scientific American*, Dec. 1964, p. 78. **368.** AP/Wide World Photos. **369.** California Institute of Technology Archives. **370.** Francis Crick. **371.** BBC Television Archives. **372.** Max Perutz, Laboratory of Molecular Biology, Medical Research Council, Cambridge, England. **373.** Alex Semenoick; from William H. Stein and Stanford Moore, "The

Chemical Structure of Proteins," *Scientific American*, Feb. 1961, pp. 83–83.   **374.** Woodson Research Center, Fondren Library of Rice University, Houston, Texas.   **375.** Ralph Morse, LIFE Magazine © 1965 Time, Inc.   **376.** The Jackson Laboratory.   **377.** Dr. Michael E. DeBakey/ Baylor College of Medicine, Houston, Texas.   **378.** National Institutes of Health.   **379.** © 1987 Fred Ward/ Black Star.   **380.** General Electric Research & Development Center.   **381.** Merck, Sharp & Dohme.   **382.** © Aileen & W. Eugene Smith/Black Star.   **383.** National Archives, from the records of the Nuremberg Trials. **384.** National Center for Jewish Films.   **385.** Vincent G. Dethier, *To Know a Fly*, San Francisco, Holden-Day, 1962, p. 38.   **386.** National Archives (NA-80G-473739).   **387.** (top) William Henry Jackson, 1873, Moraines on Clear Creek, Valley of the Arkansas, Colorado; collection: U.S. Geological Survey.   **387.** (bottom) Mark Klett and JoAnn Verburg for the Rephotographic Survey Project, 1977, Clear Creek Reservoir, Colorado.   **388.** Ted Spiegel/ Black Star.   **389.** Reprinted with permission of The Electric Power Research Institute.   **390.** Rex Features, Ltd. **391.** Rex Features, Ltd.   **392.** © Brian Bevan/Ardea London.   **393.** © 1977 Los Angeles *Times*, reprinted with permission.   **394.** AP/Wide World Photos.

Index

The numbered references in this index refer either to textual material, which is designated by page numbers in lightface, or caption material, which is designated by illustration numbers in **boldface**. It is suggested that, in addition to the page and illustration number references given here for any particular subject, the reader also consult the Picture Sources and Credits section for any illustration listed, which may provide some additional information.

Abel, John Jacob, **354**, **355**
acid rain, **388**, **389**
adaptive radiation, 113, **161**
Adélie penguin, **79**
adsorption chromatography, **85**
Agent Orange, **391**
aggression, **282**
agriculture, 243–244
   research stations, 26, **40**, **41**, **332**
   See also Lysenko, Trofim Denisovich
AIDS, **394**
Aleksandra Feodorovna (Russian empress), **70**
algae, fossil, **168**, **170**
alpha helix (protein), 170, **369**
American Museum of Natural History, 69, **101–103**
   expeditions, **93**, **97**
amino acids, 170
   creating, 181, **251**
   extraterrestrial, 182
   research apparatus, **88**, **222**
   ribonuclease, **373**
Andrews, Roy Chapman, **97**
animal behavior, 203–204, **279–290**
   influenced by industry, **392**
animal colonies, 14, 37, 159, **220**, **376**

animal language, **374**
animal rights movement, 37, 38, **60**
animals in research, 37–38
anthropology, physical, 193–194
antibiotics, 215, 216, **302**
   effects on bacteria, **303**
   production, **301**
   See also penicillin
antitoxins, 159
antivivisectionism, 38, **60**
ants, **287**
Aristotle, 38
arthritis, **224**
artificial organs, 253, **349**
   heart, 254, **359–361**
   iron lung, **350**
   kidney, **354**, **356**, **359**
   womb, **362**
aspirin, 215
atomic bomb, 281, **385**, **386**
Atwater, Wilbur Olin, **189**
Auschwitz (concentration camp), **384**
Australopithecus, 193, 194
autoradiography, 48, **72–74**
auxins, xii, **217**, **338**
auxograph, **78**
Avery, Oswald, 169, **244**

bacteria, **172**
   DNA research, 169, **244**
   fossil, **168**
   oil-eating, **380**
   See also specific bacteria
bacteriology, 3, 13
bacteriophage, 169, **233**, **245–248**
Bailey, Leonard, **358**
Bakken, Earl, **352**
Balinsky, Boris, **116**
Banting, Frederick, 215, **216**
Bateson, William, **132**
Bayliss, William, 149
Beadle, George, **139**, 170, **243**
behavioral sink phenomenon, **289**
Behring, Emil von, **48**, 215
Benedict, Francis, **191**
Berezovka mammoth, **95**, **96**
Bern University Institute of Physiology, **19**, **20**
Best, Charles, **216**
"big science," 14
biochemistry, 3, 47, 57, 58, 84, 159–160, **226**, **227**, **368**
bioinsecticides, **339**
biological feedback, 138
biologicals, 4, 215
biomedical engineering, 253–254
Biometrika, **77**

biophysical insect control, 244, **340**
birds, **4**, 38, **53**
bird's-nest orchid, **65**
birth control, 233–234, **317**, **324**
   China, **328–331**
   India, **325–327**
Blackman, Frederick Frost, **142**
blastoderm, **113**
blood, **364**
   proteins, **87**
Bock, Arlie, **192**
botany, xiii, 13
Boule, Marcellin, **270**
Boveri, Theodor, **110**
Bragg, William Henry, 169
Bragg, William Lawrence, 169
brain, **211**
   size, **272**
breeders' support for genetics, **126**, 243
breeding, **126**, **128**, 243
   See also crossbreeding; hybridization
Bridges, Calvin Blackman, **127**, **132**
Briggs, Robert W., **257**
British Museum (Natural History), **3**, **4**, 69
British tit, **392**
Brown, Barnum, **104**
Brown, Louise Jay, **249**
Brown-Séquard, C.-E., 149
Brownsville, N.Y. (Sanger clinic), **319**
Bryan, William Jennings, **266**
Bugher, John C., **308**
Burgess shale, **92**
Burnet, Macfarlane, **307**
Burnham, Charles, **139**
Burroughs Wellcome & Co., **13**, **14**

Calhoun, John, **288–290**
calorimetry, **189**, **190**
Cambrian life, **92**
Cambridge University, **26**, **27**, 169
camera lucida, **133**, **134**
cancer, **49**, **50**, 305
Cannon, Walter Bradford, **199**
Canti, R. G., **175**
Čapek, Karel, **347**
carbon cycle, **148**
Carnot, Sadi, **9**
Carrel, Alexis, **90**, **357**
carrot plants, **58**
Carson, Rachel, 244

castration, **54**
catastrophism, 70
Cavendish Laboratory (Cambridge), 169
celery, **184**
cell division, 125, **175**
cell structure, **177**
cell theory, 3, 125
Central Crops Research Station (N.C.), **41**
Central Dogma (molecular biology), 170, **238**
centrifuge, human, **194**
centrosome, **110**
Cepaea nemoralis (snail), **165**
Chain, Ernst Boris, 216, **297**
Chakrabarty, Ananda, **380**
Chance, Clinton, **320**
Chargaff, Erwin, 170
Chase, Martha, 169
chemistry. *See* biochemistry; organic chemistry; pharmaceutical industry; physical sciences
chemosynthetic organisms, 181
chemotherapy, 215, **291**
chick, 81, **107**, **113**
chickens, **54**
chimpanzees, **279**, 281
China, birth control, **328–331**
chi-square test, **77**
chlorophyll, **231**
chloroplasts, **230**, **231**
chromatography, 58, **85**, **86**, 88
chromosome map, 92, **133**, **137**
chromosomes, 91–92, **122**, **127**, **132**, **136**
   Drosophila, **130**, **131**, **133**, **134**
   human, **135**
   mouse, **137**
cilia, **180**
Clark, Barney, 254, **361**
Clear Creek (Colorado), **387**
cleavage, 81, **110**
cloning, 182
   plants, 58, **255**, **256**
coacervates, **253**
cockroaches, 38
codons, 170
coelacanth, **157**
Cole, Sidney W., **84**
Collip, James Bertram, **216**
competition for resources, 103, **147**
computer, electronic, 48, **344**
computerized tomography, 48, **75**
Comstock Law, 233
Cooley, Denton, **361**

Cooper, L. Gordon, **207**
corn, **138**, **139**, **335–337**
corn earworm, **335**, **339**
coronary bypass operations, **357**
Correns, Carl, **124**
cortisone, **223–225**
cotton, **334**, **339**
Courtenay-Latimer, Marjorie, **157**
cows, **343–345**
crabs, **80**
crayfish, **218**
cretinism, **214**
Crick, Francis Harry Compton, **1**, 169, 170, **234**, **237**
crop rotation, **332**
crossbreeding, 92, **127**
   double cross, **336**
   See also hybridization
crystals of biological substances, **171**, **213**, **223**, **234**, **298**, **366**, **367**
culture techniques, 38, 58, **59**, 90, **91**, 182
Cyclomedusa davidi (jellyfish), **169**
cyclosporine, 254
cytochrome oxygenase, **226**
cytogenetics, 92, **139**
cytoplasm, structure, **178**

Dachau (concentration camp), **383**
dairy production, **343**
Darrow, Clarence, **266**
Dart, Raymond, 193
Darwin, Charles, **6**
   evolution, 3, 69, **160**, **161**
   origin of life, 181, **250**
   plants bending toward light, xii
dawn man, **265**, **272**
Dawson, Charles, **269**
DBA (inbred mouse strain), **50**
DDT, 243, 281, **311**
DeBakey, Michael, **377**
decompression chamber, 193
Delage, Yves, 15, **100**
deoxyribonucleic acid. *See* DNA
development. *See* embryology
developmental biology, 82
DeVries, William, **361**
dextran, **91**
diabetes, 215
dialysis, 253, **355**, **356**
Dickinson, Robert Latou, 234
Dill, David Bruce, **192**
dinosaurs, 70, **100**, **104**, **105**
dioxin, **390**, **391**

diphtheria, **48**
DNA, **1**, **74**, 169–170, **232**, **233**, **238**
    as carrier of heredity, **244**, **245**
    plasmid, **188**
    replication, **239**
    structure, **234–236**, **370**
    transcription, **242**
    Watson-Crick paper, **237**
    See also nucleic acids
Dobzhansky, Theodosius, **164**
dogs, 38, **61**, **208**, **210**, **216**
Dohrn, Anton, **33**
Domagk, Gerhard, 216, **294**
double helix. *See* DNA
Drinker, Philip, **350**
Drosophila (fruit fly), 38, **51**, **52**, 91–92, **122**
    chromosomes, **130**, **133**, **134**
    head, **182**
    mutations, **131**, **141**
    X rays, **140**

East, Edward M., **138**
ecology, 103–104, **143–156**, 244
ecosystem, **142**
Ediacaran fauna, **169**
education, laboratory method, 14
    See also specific institutions
egg (ovum), **118**, **120**, **121**
Ehrlich, Paul, 4, **10**, 215, 225, **291**, **292**
Einthoven, Willem, **201**, **202**
electrification, **12**, 14, **17**, **198**
electrocardiograph, 138, **201**, **206**
electroencephalograph, 138, **365**
electron microscope, 47, **56**, **64**, **67**, 125
    comparison with optical microscope, **68**
    early micrograph, **65**
    See also shadow casting
electrophoresis, 58, **87**
elephants, **158**
Elton, Charles, 103, **143**, **144**, 146
embryology, 13, 81–82, **107–117**
embryo splitting, **346**
embryo transfer, **345**
Emerson, Rollins, **139**
endoplasmic reticulum, **178**
environment
    management of, 104, **149**, **150**, 282, **387**
    pollution, 281

enzymes, 159, 170, **226**, **227**, **243**, 366, **367**
Eocene life, **98**, **99**
Eohippus (Eocene horse), **99**, **106**
epidemics, **307**
ergotoxine, **14**
Escherichia coli (bacterium), **74**, **245**, **247**, **248**
estrogen, **176**
ethology, 203–204, **282**, **283**
eugenics, 4, 91, 234, **322**, **323**
Eugenics Record Office (New York), **129**
evolution, 3, 69–70, 113–114, **160**, **162–165**
    adaptive radiation, **161**
    cultural change, **392**
    elephant, **158**
    horse, **106**
    human, 193, **263–265**, **268**, **272**
    mutation theory, **125**
    nephron, **363**
    synthetic theory, **166**
Ewart, James Cossar, **128**
exobiology, 181
exoskeletal walking device, **351**
experimental biology, xiii, 14, 160

Faculty of Pharmacy (Paris), **28**, **29**
Falkenbach, Otto, **103**
family planning. *See* birth control
famine, 226, **306**
fermentation, 3, **7**
fertilization, **121**
fertilizer, **152**, **153**
fibrin, **185**
Fischer, Emil, 160, **226**
flagellum, **180**
Fleming, Alexander, 216, **297**, **300**
Fleming, Ambrose, **268**
flight and physiological stress, **193**, **194**
floor plans, **18**, **19**, **21**
Florey, Howard, 216, **297**
fluorescence microscopy, **69**
fluoroscope, **71**
food chain, 104
food web, **146**
foreskin, **91**
forestry, **150**, **155**, **156**
Forster, Roy, **81**
fossils
    Ediacaran, **169**
    Great Slave Lake algae, **170**

Gunflint Chert, **168**
    See also evolution; paleontology; *specific organisms*
four-o'clock, **127**
Fox, Sidney, **254**
Fraenkel-Conrat, Heinz, **55**
Frankenstein, **347**
Franklin, Rosalind, 169, **234**, **236**
freemartin, 150, **212**
Freud, Sigmund, 203, **278**
frogs, 37, **72**
fruit flies. *See Drosophila*
function, 137–138, 265
    evolution theory, 3
fungi, **154**
    spore, **181**

Galápagos Islands, 113, **161**
Galton, Francis, **129**
Gandhi, Indira, **327**
Ganong, William F., **78**
Gauze (Gause), Georgi Frantsevich, 103, **147**
geese, **282**, **283**
generation, 81
genetic engineering, 182, **379**, **381**
    benefits, **262**
    pioneering experiments, **188**, **368**
    risks, **260**, **261**, **378**, **393**
genetics, 91–92, **122–140**, **333**
genocide, 91
germ-free isolator, **89**
germs, 3, **7**
Gilbert, Joseph Henry, **40**
Gini, Corrado, **321**
glomerular capsule, **82**
Gobi Desert, **97**
Goodall, Jane, **281**
Goodlin, Robert, **362**
goosefish, **81**
gorillas, **280**
Gosling, Raymond, **234**
government support of research, 25–26
graphs, 137, 138, **204**, **207**, 265, **365**
Gregory, William King, **263**
Grinnell, Joseph, 103
Grundman, Jack, **351**
guinea pig, **48**
Gunflint Chert (Ontario), **168**

Haldane, J. B. S., **253**
Harlow, Harry, 204, **286**

Harvard Fatigue Laboratory, **192**
Harvard Medical School, **17**, **18**
HeLa cell line, 38, **57**, **59**, **259**
hemoglobin, **372**
Hench, Philip, **224**
Henderson, Lawrence Joseph, **364**
hepatitis, **381**
herbicides, **262**, **338**, **391**
Hershey, Alfred, 169
heterosis, **138**
Hiroshima, **386**
histology, 47, **174**
Hitchock, Fred, **193**
Holley, Robert W., **368**
Holzloehner, E., **383**
homeostasis, 138
hominids
    brain size, **272**
    fossils, 193–194, **269**, **270**, **273–275**
Homo erectus, 193
Homo habilis, 194
honeycreepers, **161**
Hooton, Earnest, **271**
Hopkins, Frederick Gowland, **84**
hormones, 149–150, 159
    See also specific hormones
horses. *See Eohippus*
human beings, experimentation on,
    383, **384**
Humboldt, Alexander von, 103
Hurxthal, Lewis, **192**
Huxley, Aldous, 182
Huxley, Julian, **321**, **374**
Huxley, Thomas Henry, 181
hybridization, **123**, **124**, **126**, **138**,
    243, **333**
Hyde, Ida, **173**

imaging techniques, 47–48
immunity, 225
inbred animal strains, **50**
inclusion bodies, **171**
independent assortment, law of,
    123
India, birth control, **325–327**
Indiana Jones, **97**
induction of development, **115–117**
industrial melanism, 113–114, **162**,
    163
industrialization of research, 14,
    160, 182, 275, **379**
insect control, 243–244, **334**, **335**,
    340, **341**
insecticides, 243, **311**, **312**, **339**

insulin, **86**, **215**, **216**
intercropping, 244, **341**
interferon, **91**
intrauterine device, **326**
in vitro conception, **249**
IQ studies, 193, 204
iron lung, 253, **350**
irrigation, **342**
isolation of populations, 114, **164**
isolator, germ-free, **89**
Ivanovsky, Dmitri Iosifovich, **171**

Jackson Laboratory, **376**
Jacob, François, 170, **240**
Jarvik, Robert, **361**
Java man, 193
jellyfish, fossil, **169**
Johanson, Donald, **275**
Jones, Donald F., **138**, **336**
jumping genes, 92, **139**
jungle research sites, 35, 36, **308**
Jurassic life, **100**
Just, Ernest Everett, **109**

Kaingaroa State Forest (New Zea-
    land), **156**
Kaiser Wilhelm Institute for Medical
    Research, 25, **38**
Kendall, Edward, **213**, **224**
Kendrew, John, **371**
Kettlewell, H. B. D., 114, **163**
Khorana, H. Gobind, **368**
kidney, **82**, **363**
King, Thomas J., **257**
kingdoms (organisms), 114, **167**
Koch, Ludwig, **374**
Koch, Robert, 4, **11**, 215
    identifying disease-causing micro-
        organisms, 37
Kolff, Willem, **359**, **360**
Krebs, Hans Adolf, 160, **228**
Krebs cycle, **228**
kymograph, 13, **27**, 57, 130, **197–
    199**, **209**

laboratories, 4
    containment, **378**
    growth of, xiii, 13–14, **18**, 25, 282
    See also marine laboratories;
        specific institutions

Lacaze-Duthiers, Henri de, **15**
Lack, David, **161**
Ladygina-Kots (Kohts), Nadezhda,
    279
Laika (Soviet space dog), 38, **61**
Lamarck, Chevalier de, 113, **160**
Lang, Fritz, **347**
language, animals, **280**, **374**
Larson, John A., 138
laryngoscope, 47
Lawes, John, **40**
leaf, **184**
leafhoppers, **334**
Leakey, L. S. B., 194
Leakey, Mary, 194, **273**
Leakey, Richard, **274**
lie detector, 138, **200**
Lillie, D. G., **142**
Lillie, Frank Rattray, 150, **212**
Lister, Joseph, 4
Lister Institute (London), 4
Little, Clarence, 49, **50**, **320**
lobsters, **218**
Loeb, Jacques, **111**
Loeffler, Friedrich, **172**
Lombard, Warren Plimpton, **209**
Lorenz, Konrad, 204, **282**, **283**
Lucy (fossil hominid), 194, **275**
Ludwig, Carl, **209**
lung, of bird, **183**
Lwoff, André, **240**
Lysenko, Trofim Denisovich, 113,
    243, **333**
lysogeny, **240**
lysosomes, **68**

MacLeod, Colin, 169, **244**
Macleod, John James Rickard, **216**
Macritherium, **158**
macrophage, **186**
malaria, 38, 53, 225–226, **311**, **312**
Malthus, Thomas, 233
mammoth, 70, **95**, **96**
Mangold, Hilde, **115**
manometric apparatus, **228**
Marine Biological Laboratory
    (Woods Hole), 25, **30–32**
marine laboratories, 13, 25, 81
    See also specific institutions
Marton, L., **65**
mass extinctions, 70
mastodon, **97**
mathematics, **77**
McCarty, Maclyn, 169, **244**

McClintock, Barbara, 92, **139**
McCollum, Elmer Verner, **220**
Mead, Margaret, **276**
mechanistic interpretation of life,
    111
medicine, 215–216, 225–226, 253–
    254
    monitoring, use of, 138, **205**, **206**
    nineteenth century, 4, 215
Mendel, Gregor, 91, **123**, **124**
Merck Sharp & Dohme, **381**
Mercury (U.S. space program), 62,
    207
mercury poisoning, **382**
Mesozoic life, **103**, **104**
metabolism, 160
    agriculture studies, **343**
    research apparatus, 137, **189–192**,
        228, **229**
Metchnikoff, Élie, 4, **10**, **16**, 225
Metropolis (film), **347**
mice, 37–38, **49**, **50**, 258, 290
    chromosome map, **137**
microbalance, xii
microbiology, 3, 13
microcinematography, **175**
microdissection, 57
microevolution, 114, **165**
micropuncture sampling, 82
microscope, 47, **66**, **76**
    staining techniques, 125, **174**
    See also electron microscope; fluo-
        rescence microscopy
microsomal particles, **178**
Miller, Stanley, 181, **251**, **252**
Minamata Bay (Japan), **382**
mitochondria, **68**
mitosis, **136**
    See also cell division
molecular biology, 160, 169–170,
    232–242, 266, **368**
    molecular models, **1**, **369–372**
molybdenum, **151**, **153**
Monera, **167**
monkeys, 38, 62
Monod, Jacques, 170, **240**
Monsanto Life Sciences Research
    Center, **379**
Morgan, Thomas Hunt, 91–92, **122**,
    127, **131**, **132**, **158**
mosquitoes, 53
Mosso, Angelo, **209**
mother-infant bond, **286**
moxalactam, **303**
Muirhead, Hilary, **372**
Muller, Hermann J., **52**, 92, **140**

Murchison meteorite, 182
muscle striations, **179**
museums, natural history, 3, 69, 70
    See also specific institutions
mutations, 92, **125**, **131**, **140**, **141**,
mycorrhizae, **154**
myoglobin, **371**
myograph, 149

Naples Zoological Station (Italy), 13,
    25, **33**, **34**
National Institute for Medical Re-
    search (U.K.), 39
National Museum of Natural His-
    tory (Paris), **2**
natural history, xiii, 3, 13, 103
Natural History Museum (Berlin),
    5
natural selection, 69, 103, 113, **160**,
    166, 287
nature versus nurture, 91
Nazi Germany, 91, 193, 281, **323**,
    383, **384**
Neanderthal man, **270**
neo-Darwinians, 113, 243
neo-Lamarckians, 113, 243, **333**
nephron, **363**
neurohormones, **218**
neurophysiology, 149, **208**
neurosciences, xiv
Neurospora crassa (red bread mold),
    169, **243**
neurotransmitters, 150
newt, **114**
niche, ecological, 103, **147**
Nirenberg, Marshall W., **368**
nomogram, **364**
nucleic acids, 91, **238**
    See also DNA; RNA
nutrition. *See* metabolism
Nuttall, George H. F., **89**

octopamine, **218**
Oenothera lamarkiana (evening prim-
    rose), **125**
Olduvai Gorge, 193, **273**
Oliver, C. P., **52**
onion root cells, **136**
Oparin, Aleksandr Ivanovich, 181,
    253
operating microscope, **76**
operon theory, 170, **240**, **241**

ophthalmoscope, 47
oral contraceptive pill, **326**
organ bank, **375**
organic chemistry, 3, **221**, **373**
organizer substance, **115**, **117**
organotherapy, 215
origins of life, 181
    Darwin, 181, **250**
    Fox, **254**
    Haldane, **253**
    Miller, **251**
    Oparin, **253**
    Urey, **252**
Osborn, Henry Fairfield, **101**, **265**,
    272
oscilloscope, 138, **202**, **206**
Osler, William, **214**
overgrazing, **149**
ovulation, **118**
oxyhemoglobin, **372**

pacemaker, 254, **352**, **353**
Painter, Theophilus Shickel, **52**,
    133, **134**
Palade, George, **178**
Palaeosyops, **101**
paleontology, 69–70, **92–105**
Paluxy River (Texas), **100**
pancreas, 68
paper chromatography, **86**
parakeet, **195**
paramecia, **147**
Park, Thomas, 103
parthenogenesis, **111**
Pasteur, Louis, 3, 4, **9**, **89**
    disease agents, **8**
    fermentation, **7**
Pasteur Institute, 4, **16**, **25**, 37
patenting of organisms, **380**
Patterson, Francine, **280**
Pauling, Linus, 170, **369**
Pavlov, Ivan Pavlovich, 149, **208**
Pearl, Raymond, **321**
Pearson, Karl, 77
peat moor, **144**
Penfield, Wilder, **211**
penicillin, 60, 216, 297, **298**
    production, **299**, **300**
peppered moths. *See* industrial me-
    lanism
Perutz, Max, **372**
petrified forest, **94**
petunias, **262**
phagocytosis, **16**

pharmaceutical industry, 4, **60,
    223**
    See also specific companies
photography, 48, 125
    ethnography, **276**
    microcinematography, **175**
photosynthesis, **83**
    See also chloroplasts
physical sciences, influence on biol-
    ogy, 3, 4, 14, 57, 58, 160, 265,
    266
physiology, 57
    instruments and apparatus, 13,
        137, 138, **197, 344**
    of work, 137
    plants, **23, 78**
    See also hormones; metabolism;
        neurophysiology; *specific instru-
        ments*
Piltdown man, 193, **269**
plant-animal cell (hybrid), **259**
plant physiology, **23, 78**
plasmid, **188**
Platyklodon, **97**
plethysmography, **196**
Pluteus II (research vessel), **44**
polio, 225, **313, 314**
Polley, Howard, **224**
pollution, **382**
ponds, **145**
population biology, 104
population dynamics, animal stud-
    ies, **288–290**
potato plants, **255, 256**
Precambrian life, 113, **168–170**
predation, 103
primrose. *See Oenothera lamarkiana*
prontosil, 216, **294**
proteins, 169
    alpha helix, **369**
    hemoglobin, **372**
    myoglobin, **371**
    synthesizing, 170, **232, 238,
        240**
Protista, **167**
protoplasm, 181
protoplasts, **255**
protozoa, **112**
Pseudomonas (bacterium), **303,
    380**
psychoanalysis, 203, **278**
psychology, 203, 204, **277**
public health campaigns, 225–
    226, **304–306, 311–316**
pyramid of numbers (in
    ecosystems), **146**

rabbits, **60, 118**
race and racism, 193, 194, **265, 271,
    323**
radiation
    effects, 281
    mutations, 92, **140, 141**
radioactive tracers, 48, 58, **83,** 160,
    169, **229, 230**
radio monitoring, **79**
rain forest, **146**
Rascher, Sigmund, **383**
rats, 38, **284, 288, 289**
rat universe, **288, 289**
reclamation, **152, 153, 342**
Recombivax HB, **381**
recording instruments, 57
    See also physiology, instruments
        and apparatus; *specific instru-
        ments*
red blood cell, **185**
reductionism, 266
reflexes, 149, **209, 210**
    conditioned, **208**
    pathways, **219**
regeneration, **112**
Reichstein, Tadeus, **224**
replica (electron microscopy),
    181
reproductive biology, 82
reproductive technology, 182,
    249
    agriculture, 244, **345, 346**
    cloning, 58, 182, **255, 256**
research institutes, 4, 25
respiration
    analysis, **191, 195**
    bird lung, **183**
    calorimetry, **189, 190**
    plethysmography, **196**
rhesus monkeys, **286**
Rhoades, Marcus, **139**
ribonuclease, **373**
ribonucleic acid. *See* RNA
ribosomes, **68,** 178, **231**
Richards, Alfred Newton, **82**
Ricketts, Edward, **43**
RNA, **73, 238, 240, 242**
robots, 253, **347, 348**
Rockefeller Foundation, 226
Röntgen (Roentgen), Wilhelm Kon-
    rad, **70**
roots, enhanced, **154**
Rosa, Edward Bennett, **189**
Roscoff Biological Station (France),
    13, **15,** 26, 42, **44,** 108
Ross, Ronald, 38, **53**

Rothamsted Experimental Station,
    40
Ruska, Ernst, **64**

salamander, **116, 117**
Salk, Jonas, **314**
Salvarsan, 215, **291, 292**
Samoa, **276**
sand dunes, **143**
Sanger, Margaret, 233–234, **318, 319**
scanning electron microscope, 126,
    184
Scarritt Expedition, **93**
Schleiden, Mathias, 3
Schull, George Harrison, **336**
Schwann, Theodor, 3
Scopes, John T., 193, **267**
Scopes trial, 266
Seatopia (Japanese undersea habi-
    tat), **46**
sea urchin, 38, **110, 120**
secretin, 149
segregation, law of, **123**
Seireg, Ali, **351**
serotonin, **218**
serum therapy, **48,** 215
Seveso (Italy), **390**
sex hormones, **54,** 149–150, **212**
shadow casting (electron micros-
    copy), 47, **67,** 181
Shelford, Victor Ernest, 103, **145**
Shelley, Mary Wollstonecraft, **347**
Sherrington, Charles Scott, 149, **210**
Shull, George Harrison, **138**
Siemens and Halske Company, **64**
Simpson, George Gaylord, **98**
Skinner, B. F., 204, **284**
Skinner box, 204, **284**
slavery, **287**
sliding filament theory, **179**
Sloan, John French, **71**
Slocumb, Charles H., **224**
Slye, Maud, **49**
smallpox, 225, **315, 316**
Smith, J. L. B., **157**
Smith, W. Eugene, **382**
Smith College (Mass.), **21–23, 78**
smoking, campaigns against, **304**
social goals of science, 282
sociobiology, 204, **287**
soil
    fungi, **154**
    infertile, **151, 152**
    reclamation, **153**

Sorbonne (Paris), **24**
Soviet Union, agriculture. *See* Lysenko, Trofim Denisovich
Spacelab, **47**
space medicine, 47, **61**, **62**, **194**, **207**
Spemann, Hans, 82, **114**, **115**, **117**
spermatozoa, **119–121**
Spillman, William Jasper, **126**
Sprigg, Reginald, **169**
squid, 38
staining (microscopy), 125, **174**
Stanley, Wendell, **55**
Starling, Ernest, 149
starlings, **285**
statistics, **77**
Steinbeck, John, **43**
Stentor, **112**
Steward, Frederick, **58**
Stone, W. S., **52**
Stopes, Marie, 233, **317**
streptomycin, **301**
string galvanometer, **17**, **202**
stromatolites, **170**
Sturtevant, Alfred Henry, **127**, **132**
sulfa drugs, 216, **294–296**
superovulation, **345**
surgery, **357**, **377**
Svedberg, The, 58
synthetic theory of evolution, 70, 113, 114, **166**
syphilis, 215, 216, **291–293**

tadpoles, **257**
Tansley, Arthur G., **142**
Tatum, Edward L., **139**, 170, **243**
Tektite (U.S. undersea habitat), **45**
telegony, **128**
Tesla, Nikola, **12**
test-tube babies, **249**
Tetrabelodon, **158**
Texas, University of (Austin), **52**
Theiler, Max, **308**
thermal proteinoid, **254**
Thierfelder, H., **89**
Thimann, Kenneth V., xii
Thomashuxleya externa, **98**
thyroxine, **213**, **214**
Tiselius, Arne, 58, **87**
titanothere, **101**

tobacco mosaic disease, **171**
tobacco plants, **55**
"transforming principle" (DNA), 169, **244**
transgenic organisms, **275**
transmission electron microscope, **126**
transplantation
    blood vessels, **357**
    cell nuclei, 182, **257–259**
    heart, **358**
    interspecies, **358**
    limb, **116**
    organs, **254**
    tissue, **115**
Triassic life, **103**
trypsin, **366**, **367**
trypsinogen, **366**
tryptophan, **84**
Tschermak von Seysenegg, Erich, **124**
Tsvet (Tswett), Mikhail Semenovich, 58, **85**
tuberculosis, **11**
Turin (Italy) exposition, **14**
turtle, **103**
2,4-D (herbicide), **338**
Tyrannosaurus rex, **104**

ultracentrifuge, 58
undersea habitats, **45**, **46**
urease, **366**
Urey, Harold, **252**
uterus, response to estrogen, **176**

vaccines, 225, 226, **314**
    genetically engineered, **381**
Van Slyke, Donald, **222**
vasectomy, **327**
Vavilov, Nikolai Ivanovich, **333**
vernalization, **333**
Vietnam war, xii, **391**
Virchow, Rudolf, 3
viruses, 47, **56**
    tobacco mosaic, **55**, **171**, **187**
    See also bacteriophage
vitamins, **84**, 159, **220**

vividiffusion, **354**, **355**
vivisection, 37, 57
Vries, Hugo de, **124**, **125**

Walcott, Charles, **92**
walking machine, **351**
Warburg, Otto, 160, **226–228**
Wassermann test, **293**
Watson, James Dewey, **1**, 169, 170, **234**, **237**
Watts, Newman, **268**
Wearn, Joseph, **82**
wellness, 226
Went, Frits, xii, **217**
Wernicke, Erich, **48**
whales, **5**
wheat, **126**, **333**
White, Philip R., **58**
white blood cell, **186**
Wieland, Heinrich, **221**
Wilkins, Maurice, 169, **235**
Williams, Robley, **67**
Williams, Timothy J., **346**
Wilson, Edward, **287**
Woodward, Arthur Smith, **269**
World Health Organization, 226, **311**, **312**, **315**, **316**
World Population Conference (1927), 234, **320**, **321**
World's Columbian Exposition, **12**

X-ray diffraction crystallography, 169, **234**, **235**
X rays, **47–48**, **70**, **71**

yeast, **7**
yellow fever, 226, **308–310**
    research sites, **35**, **36**

zebras, **128**
Zinjanthropus, **194**, **273**
zonation of vegetation, **143**, **144**
zoology, xiii, 13